AF598069

Mesenchymal Stem Cells

METHODS IN MOLECULAR BIOLOGY™

John M. Walker, SERIES EDITOR

460. **Essential Concepts in Toxicogenomics,** edited by Donna L. Mendrick and William B. Mattes, 2008
459. **Prion Protein Protocols,** edited by Andrew F. Hill, 2008
458. **Artificial Neural Networks:** Methods and Applications, edited by David S. Livingstone, 2008
457. **Membrane Trafficking,** edited by Ales Vancura, 2008
456. **Adipose Tissue Protocols,** Second Edition, edited by Kaiping Yang, 2008
455. **Osteoporosis,** edited by Jennifer J. Westendorf, 2008
454. **SARS- and Other Coronaviruses:** Laboratory Protocols, edited by Dave Cavanagh, 2008
453. **Bioinformatics,** Volume 2: Structure, Function, and Applications, edited by Jonathan M. Keith, 2008
452. **Bioinformatics,** Volume 1: Data, Sequence Analysis, and Evolution, edited by Jonathan M. Keith, 2008
451. **Plant Virology Protocols:** From Viral Sequence to Protein Function, edited by Gary Foster, Elisabeth Johansen, Yiguo Hong, and Peter Nagy, 2008
450. **Germline Stem Cells,** edited by Steven X. Hou and Shree Ram Singh, 2008
449. **Mesenchymal Stem Cells:** *Methods and Protocols*, edited by Darwin J. Prockop, Donald G. Phinney, and Bruce A. Bunnell, 2008
448. **Pharmacogenomics in Drug Discovery and Development,** edited by Qing Yan, 2008
447. **Alcohol: Methods and Protocols,** edited by Laura E. Nagy, 2008
446. **Post-translational Modification of Proteins:** Tools for Functional Proteomics, Second Edition, edited by Christoph Kannicht, 2008
445. **Autophagosome and Phagosome,** edited by Vojo Deretic, 2008
444. **Prenatal Diagnosis,** edited by Sinhue Hahn and Laird G. Jackson, 2008
443. **Molecular Modeling of Proteins,** edited by Andreas Kukol, 2008
442. **RNAi: Design and Application,** edited by Sailen Barik, 2008
441. **Tissue Proteomics: Pathways, Biomarkers, and Drug Discovery,** edited by Brian Liu, 2008
440. **Exocytosis and Endocytosis,** edited by Andrei I. Ivanov, 2008
439. **Genomics Protocols,** Second Edition, edited by Mike Starkey and Ramnanth Elaswarapu, 2008
438. **Neural Stem Cells:** *Methods and Protocols*, Second Edition, edited by Leslie P. Weiner, 2008
437. **Drug Delivery Systems**, edited by Kewal K. Jain, 2008
436. **Avian Influenza Virus**, edited by Erica Spackman, 2008
435. **Chromosomal Mutagenesis**, edited by Greg Davis and Kevin J. Kayser, 2008
434. **Gene Therapy Protocols:** Volume 2: Design and Characterization of Gene Transfer Vectors, edited by Joseph M. LeDoux, 2008
433. **Gene Therapy Protocols:** Volume 1: Production and In Vivo Applications of Gene Transfer Vectors, edited by Joseph M. LeDoux, 2008
432. **Organelle Proteomics**, edited by Delphine Pflieger and Jean Rossier, 2008
431. **Bacterial Pathogenesis:** *Methods and Protocols*, edited by Frank DeLeo and Michael Otto, 2008

Mesenchymal Stem Cells

Methods and Protocols

Edited by

Darwin J. Prockop
Tulane University Health Sciences Center,
New Orleans, LA.

Donald G. Phinney
Tulane University Health Sciences Center,
New Orleans, LA.

Bruce A. Bunnell
Tulane University Health Sciences Center,
New Orleans, LA.

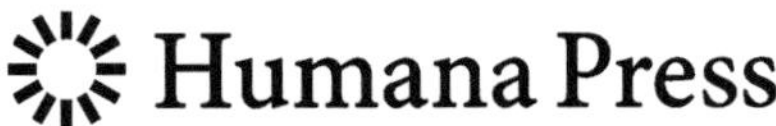

Editors
Darwin J. Prockop
Center for Gene Therapy
Tulane University
Health Sciences Center
New Orleans, LA
dprocko@tulane.edu

Donald G. Phinney
Center for Gene Therapy
Tulane University
Health Sciences Center
New Orleans, LA
dphinne@tulane.edu

Bruce A. Bunnell
Center for Gene Therapy
Tulane University Health Sciences Center
New Orleans, LA
bbunnell@tulane.edu

Series Editor
John M. Walker
Professor Emeritus
School of Life Sciences
University of Hertfordshire
Hatfield
Hertfordshire AL10 9AB, UK

ISBN: 978-1-58829-771-6 e-ISBN: 978-1-60327-169-1
DOI: 10.1007/978-1-60327-169-1

Library of Congress Control Number: 2007938053

Cover illustration: Fig.1 of Chapter 2 by Mark F. Pittenger

Printed on acid-free paper

9 8 7 6 5 4 3 2 1

springer.com

Preface

The discovery of mesenchymal stem cells is credited with Alexander Friedenstein and associates, who over 40 years ago demonstrated that pieces of bone marrow transplanted under the renal capsule of mice formed a heterotopic osseous tissue that was self-maintaining, self-renewing, and capable of supporting host cell hematopoiesis. Furthermore, Friedenstein showed that the osseous-forming activity of bone marrow was contained within the fibroblastoid cell fraction isolated by preferential attachment to tissue culture plastic. These findings confirmed that bone marrow contained separable stem cell populations capable of generating hematopoietic and connective tissue cell lineages. These studies also demonstrated that marrow-derived, plastic adherent fibroblastic (stromal) cells were capable of supporting the growth and differentiation of various hematopoietic cell types. These cells were then used as feeder layers to establish long-term bone marrow cultures in vitro, which fostered a wealth of new knowledge regarding the molecular mechanisms regulating hematopoiesis.

In the decades following Friedenstein's seminal publications, various groups labored to delineate the biological nature and differentiation potential of plastic adherent cells from bone marrow. These efforts revealed much information about their cell surface phenotype, proliferative and differentiation potential and culminated in the demonstration that clonally derived murine and human populations were multipotent, capable of differentiating into adipocytes, chondrocytes, osteoblasts, and hematopoiesis-supporting stromal cells. The latter findings confirmed the existence of a stem cell in marrow capable of generating most connective tissue cell types. Consequently, the marrow-derived, plastic adherent cells first referred to as colony-forming unit fibroblast (CFU-F) by Friedenstein, then in the hematological literature as marrow stromal cells, subsequently became known as mesenchymal stem cells. Recently, a committee from the International Society of Cell Therapy has adopted the term multipotent mesenchymal stromal cells (MSCs) to define these cells owing to the fact that a definitive description of the bona fide mesenchymal stem cell and the molecular mechanisms that regulate its self-renewal versus differentiation remain forthcoming. The literature has been confused by the frequency with which the different names for essentially the same cells have been used (*see* Fig. 1). In this compendium, the terms CFU-F, marrow stromal cell, mesenchymal stem cell, and multipotent mesenchymal stromal cell

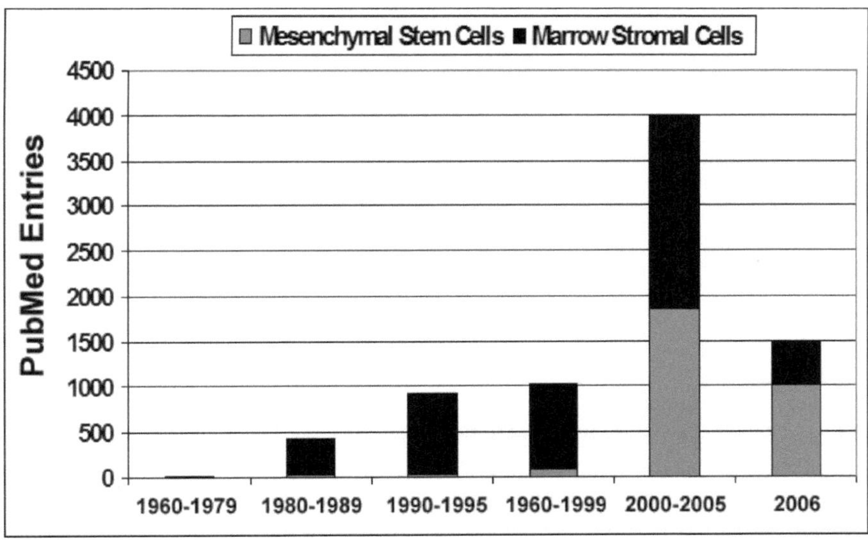

Fig. 1 Illustrated is the number of citations found in the PubMed database that contain the phrase "marrow stromal cells" or "mesenchymal stem cells" in their title or abstract over various time periods.

are deemed equivalent and herein will refer to the plastic adherent, fibroblastoid cells from marrow that are defined functionally by their capacity to undergo multi-lineage differentiation into connective tissue cell lineages.

In recent years MSCs have garnered much attention owing to their broad therapeutic efficacy. Initially, MSC administration to children afflicted with osteogenesis imperfecta was found to have a significant positive impact by reducing the severity of the disease. Promising results were subsequently reported using MSCs or related cells from bone marrow in the treatment of Hurler's syndrome, metachromatic leukodystrophy, graft versus/host disease and to enhance engraftment of heterologous bone marrow transplants. Most recently, MSCs have been shown to afford a therapeutic benefit in the treatment of myocardial infarction, stroke, lung diseases, spinal cord injury, and other neurological disorders. These results, together with the fact that MSCs can be readily isolated from small volume bone marrow aspirates, expanded to large numbers ex vivo and engineered genetically have made them extremely attractive as therapeutic cellular vectors.

Despite these advances, it has been difficult to assess the overall therapeutic use of MSCs owing to conflicting reports in the literature regarding their engraftment levels in tissues in vivo, their overall differentiation potential in vitro and in vivo, as well as their therapeutic efficacy in disease models. Although some of these discrepancies are related to limitations associated with experimental methodologies, critical differences in the preparation and expansion of donor cells used for the experiments certainly contribute to this problem, as well. Consequently, the necessity of developing standardized methods to isolate, phenotype, and evaluate the quality of MSCs is ever increasing. Accordingly, the following compendium provides detailed methodologies for the isolation and characterization of human and rodent MSCs contributed by a group of assembled leaders in the field.

Darwin J. Prockop, MD, PhD
Donald G. Phinney, PhD
Bruce A. Bunnell, PhD

Contents

Contributors

Zeni Alfonso, PhD.
Cytori Therapeutics Inc., San Diego, CA.

Bruce A. Bunnell, PhD.
Center for Gene Therapy and Department of Pharmacology,
Tulane University Health Sciences Center,
New Orleans, LA.

Stacy Carling, MS.
Pennington Biomedical Research Center, Baton Rouge, LA.

Severine G. Dubois, PhD.
Pennington Biomedical Research Center, Baton Rouge, LA.

John K. Fraser, PhD.
Cytori Therapeutics Inc., San Diego, CA.

Elizabeth Z. Floyd, PhD.
Pennington Biomedical Research Center, Baton Rouge, LA.

Jeffrey M. Gimble, MD, PhD.
Pennington Biomedical Research Center, Baton Rouge, LA.

Stan Gronthos, MSc, PhD.
Mesenchymal Stem Cell Group, Division of Haematology,
Institute of Medical and Veterinary Science, Hanson Institute,
University of Adelaide, SA, Australia.

Yuan Di C. Halvorsen, PhD.
Curagen Corp., Branford, CT.

Reza Izadpanah, DVM, PhD.
Center for Gene Therapy and Tulane National Primate Research Center,
Tulane University Health Sciences Center, New Orleans, LA.

Gail Kilroy, BS.
Pennington Biomedical Research Center, Baton Rouge, LA.

Donald G. Phinney, PhD.
Center for Gene Therapy and Department of Microbiology and Immunology,
Tulane University Health Sciences Center, New Orleans, LA.

Mark F. Pittenger, PhD.
Visiting Professor, Cardiology Dept of Medicine,
Johns Hopkins School of Medicine, Baltimore, MD.

Radhika Pochampally, PhD.
Center for Gene Therapy and Department of Pharmacology,
Tulane University Health Sciences Center, New Orleans, LA.

Darwin J. Prockop, MD, PhD.
Center for Gene Therapy and Department of Biochemistry,
Tulane University Health Sciences Center, New Orleans, LA.

Eric Ravussin, PhD.
Pennington Biomedical Research Center, Baton Rouge, LA.

Roxanne L. Reger, MS.
Center for Gene Therapy, Tulane University Health Sciences Center,
New Orleans, LA.

Beerelli Seshi, MD.
Geffen School of Medicine at UCLA, Head, Hematopathology,
Los Angeles Biomedical Research Institute at Harbor-UCLA Medical Center,
Torrance, CA.

William Swaney, MS.
Center for Gene Therapy, Tulane University Health Sciences Center,
New Orleans, LA.

Alan H. Tucker, BS.
Center for Gene Therapy, Tulane University Health Sciences Center,
New Orleans, LA.

Margaret R. Wolfe, MS.
Center for Gene Therapy, Tulane University Health Sciences Center, New Orleans, LA.

Xiying Wu, MD.
Pennington Biomedical Research Center, Baton Rouge, LA.

Isabella Wulur, MS.
Cytori Therapeutics Inc. San Diego, CA.

Joni Ylöstalo, MS.
Center for Gene Therapy, Tulane University Health Sciences Center, New Orleans, LA.

Andrew C. W. Zannettino, PhD.
Myeloma and Mesenchymal Research Group, Division of Haematology, Institute of Medical and Veterinary Science, Hanson Institute, Adelaide, SA, Australia.

Min Zhu, MD.
Cytori Therapeutics Inc. San Diego, CA.

Sanjin Zvonic, PhD.
Pennington Biomedical Research Center, Baton Rouge, LA.

Color Plates

Color Plates follow p. 92

COLOR PLATE 1	(**A**) Crystal violet stained plates of CFU-F assays performed on two different donors *(10)*. (**B**) Sc-CFU is more sensitive and reproducible than the traditional CFU assay: A: Sc-CFU assay of MSCs initially plated at varying densities and incubated for 10–11 d (mean+/–SD, n=2). B: Standard CFU assay of MSCs plated at same densities and incubated for 13–14 d (mean+/–SD, *n* = 3 or 4)
COLOR PLATE 2	The most clonal cells are characteristically FS^{lo}/SS^{lo}: A: Sc-CFU assay of cells suggesting above 80% colony formation by cells that are low in granularity and size. Crystal Violet Representative microtiter plate in which single FS^{lo}/SS^{lo} cells were deposited. The plate shown in was stained with crystal violet to count colonies after incubation in standard medium for 12–14 d
COLOR PLATE 3	Mineral deposition by MSCs cultured in osteogenic medium indicating early stages of bone formation. Stained with alizarin red S. Mag: 20X
COLOR PLATE 4	Fat globules seen in MSC culture grown in adipogenic medium indicating differentiating into adipocytes. Stained with oil red O. Mag: 20X
COLOR PLATE 5	MSC micromass pellet, grown in chondrogenic medium and stained with Toluidine Blue Na Borate. Mag: 10X
COLOR PLATE 6	Hierarchial clustering of the genes. Differentially expressed genes (as in Fig. 10.1) for Ch-6 experiment were used to cluster genes hierarchically in dChip program. The red color represents expression level above mean expression of a gene across all samples, the white color represents expression at the mean level, and the blue color represents expression lower than the mean. Co-expressed genes were defined from the clustering

picture and average profiles are shown (identified as A-J for Ch-6) on the left of the clustering picture. In the picture a row represents a gene and each column represents a sample from the time course (Day 0, 1, 7, 14, and 21)

COLOR PLATE 7 Purification and differentiation of IDmMSCs. The top panel illustrates how immunodepletion removes contaminating hematopoietic lineages from plastic adherent cultures elaborated from murine bone marrow. Images are Geimsa stained plastic adherent populations before and following immunodepletion. The bottom panel shows the potential of IDmMSCs to differentiate into chondrocytes, adipocytes, osteoblasts, myoblasts and hematopoiesis-supporting stroma when cultured under the appropriate conditions in vitro

Part I
Isolation of Human Multipotential Stromal Cells from Bone Marrow and Adipose Tissue

Chapter 1
Isolation and Culture of Bone Marrow-Derived Human Multipotent Stromal Cells (hMSCs)

Margaret Wolfe, Radhika Pochampally, William Swaney, and Roxanne L. Reger

Abstract We have developed protocols whereby a total of 30–90 × 10^6 hMSCs with an average viability greater than 90% can be produced in a single multilevel Cell Factory from a relatively small (1–3 mL) bone marrow aspirate in 14–20 d. It is possible to generate as many as 5 × 10^8 multipotent stromal cells (MSCs) from a single sample, depending on the number of Cell Factories seeded from the initial isolated hMSCs. Briefly, mononuclear cells are collected from a bone marrow aspirate by density gradient centrifugation. The cells are cultured overnight and the adherent cells are allowed to attach to the flask. Nonadherent cells are removed and the culture expanded for 7–10 d with periodic feeding of the cells. The cells are then harvested and seeded at low density (60–100 cells/cm^2) into Nunc Cell Factories. The cells are allowed to expand for an additional 7–10 d, and are then harvested.

Keywords Mesenchymal stem cells; MSCs; isolation; culture; multipotent.

1 Introduction

Recently there has been interest in developing cell and gene therapies with adult stem/progenitors cells from human bone marrow that are plastic adherent and were originally referred to as fibroblastoid colony forming units, then in the hematological literature as marrow stromal cells, subsequently as mesenchymal stem cells, and most recently as multipotent mesenchymal stromal cells (MSCs). Human MSCs (hMSCs) are appealing as a basis of cell therapies because they are: (a) readily isolated from the patient, (b) able to be expanded rapidly in vitro, (c) amenable to being genetically engineered to introduce genes of interest for therapy, (d) nontumorigenic unless subject to stress or expanded extensively in culture, and (e) part of a natural repair system, whereby they home to sites of tissue injury and repair the tissues by several mechanisms. These mechanisms include: (a) differentiation into tissue-specific phenotypes, (b) secretion of chemokines to enhance repair of damaged cells and stimulate proliferation of tissue endogenous stem cells, and (c) perhaps by transfer of mitochondria or mitochondrial DNA *(1–5)*.

From: *Methods in Molecular Biology, vol. 449, Mesenchymal Stem Cells: Methods and Protocols*
Edited by D.J. Prockop, D.G. Phinney, and B.A. Bunnell © Humana Press, Totowa, NJ

hMSCs are readily isolated from bone marrow by their adherence to tissue culture plastic and can be expanded through multiple passages in medium containing high concentrations of fetal calf serum (FCS) *(1,6–12)*. However, the proliferation rates and other properties of the cells gradually change during expansion, and therefore it is advisable to not expand them beyond 4 or 5 passages. The cells are readily cloned as single-cell derived colonies, but both the colonies and the cells within a colony are heterogeneous in morphology, rates of proliferation, and efficacy with which they differentiate. Two morphologically distinct cell phenotypes are seen in early passage, low density cultures: small, spindle-shaped cells that are rapidly self-renewing (RS cells) and large, flat cells that replicate slowly and appear more mature *(13,14)*. The proportion of RS cells remains high for several passages if the cultures are maintained at low density, but the larger cells predominate in later passages *(13–16)*.

Cultures enriched for RS cells are obtained if early passage cultures expanded to about 70% are lifted and replated at low density. However, if the cultures are allowed to expand to confluence, RS cells are not recovered on replating at low density.

MSCs differentiate into osteoblasts, chondrocytes, adipocytes, myocytes, cardiomyocytes, and other cell types *(17–21)*. They can be differentiated in vitro into skeletal muscle and cardiomyocytes *(21–25)*. Several investigators *(13,16,26,27)* identified distinctive surface epitopes on hMSCs, but none have been shown to distinguish early progenitors from more mature cells in cultures of hMSCs.

The first step is to isolate mononucleated cells from a marrow aspirate by centrifuging the sample on a density gradient and then recovering the cells that adhere to tissue culture plastic (passage zero cells). The passage zero cells are then expanded by plating them at a low density in a multilayered tissue culture flask (Cell Factory; Nunc). The expansion of the cultures at low density enhances their content of spindle-shaped and rapidly expanding cells (rapidly self-renewing or RS cells) that are replaced by large, flat and mature hMSCs if the cells are plated at higher density or the cultures are passaged more than 4–6 times. The mature hMSCs expand slowly and have less potential for differentiation than RS but retain the ability to differentiate into mineralizing cells and secrete factors that enhance growth of hematopoietic stem cells and perhaps other cells.

2 Materials

2.1 Reagents

1. α-MEM with L-glutamine, without ribonucleosides or deoxyribonucleosides (Invitrogen; catalog # 12561-049 (1 L) or -056 [500 mL]) or similar.
2. Fetal Bovine Serum (FBS) (Atlanta Biologicals Premium Select FBS, catalog # S11550) *see* **Note 1**.
3. L-Glutamine, 200 m*M* (Invitrogen; catalog # 25030-081).
4. Penicillin G (10,000 units/mL), and streptomycin sulfate (10,000 μg/mL) in solution of 0.85% NaCl (Invitrogen/GIBCO; catalog # 15140-122) *(Optional)*.

5. Hank's Balanced Salt Solution (HBSS) w/o Ca^{++} & Mg^{++} (Invitrogen; catalog # 1470-112) or similar.
6. Ficoll-Paque (Amersham Biosciences, catalog # 17-440-02) or similar.
7. Phosphate buffered saline (PBS), without Ca^{++} or Mg^{++}, pH 7.4 (Invitrogen; catalog # 10010-031) or similar.
8. Trypsin (0.25%)-EDTA.4NA (0.38 g/L) in HBSS (Invitrogen; catalog # 25200-056).
9. Trypan blue, 0.4% (Invitrogen; catalog# 15250-061) or similar.

2.2 Equipment

2.2.1 Common Equipment

1. Biological Safety Cabinet, Class II (BSC).
2. Centrifuge with swinging bucket rotor, capable of holding various tube sizes up to 200 mL.
3. Incubator, water jacketed and humidified with 5% CO_2, maintained at 37 °C.
4. Microscope, inverted, phase with a super long working distance condenser (SLWDC), *see* **Note 2**.
5. Vacuum aspiration source with tubing and waste container.
6. 37°C Waterbath.

2.2.2 Equipment for Closed System Only

1. TSCD Sterile Tubing Welder (Terumo, model SC-201A)
2. Hand-held Tube Sealer (Sebra, model 2380)

2.3 Supplies

2.3.1 Common Supplies

1. Electric or manual pipet filler/dispenser for mouth-free pipeting of solutions (0.1 to 25 mL).
2. Single channel pipetors, air displacement, capable of accurately measuring from 10 µL to 1000 µL, i.e., Eppendorf Research Series 2100 or similar.
3. Sterile aerosol barrier pipet tips, 10, 20, 200, and 1000 µL.
4. Hemocytometer with cover slip.
5. 15, 50, and 200 mL sterile plastic disposable conical centrifuge tubes.
6. Plastic disposable snap-cap centrifuge tubes: 1.5 mL.
7. 1 L Nalgene bottles (Nalge, catalog # 2006-0032) or other 1 L containers that are sterile or can be sterilized. They are used to collect cell harvest from Cell Factory.

8. Sterile serological pipets: 1, 2, 5, 10, 25, and 50 mL.
9. Vacutainer tubes w/sodium heparin (B-D; catalog # 366480) or similar.
10. 175 cm^2 tissue culture flasks (Nunc; catalog # 159910) or similar, *see* **Note 3**.
11. Sterile plastic Pasteur transfer pipets.
12. Sterile pipets for aspiration source.
13. Sterile 250 mL filter units 0.22 μm pore size (Millipore, Stericup, catalog # CGPUO2RE).
14. Sterile 500 mL filter units, 0.22 μm pores (Millipore, Stericup, catalog # SCGPU05PE).
15. Sterile 1000 mL filter units, 0.22 μm pores (Millipore, Stericup, catalog # SCGPU11PE), *see* **Note 4**.
16. 10-tray Cell Factory, 6,320 cm^2 culture area (Nunc, catalog # 170009, case of 6; catalog # 164327, case of 2).

2.3.2 Supplies for Open System Only

1. Cell Factory Start-up kit, Sterile, Nunc catalog # 170769 (contains the following components: (2) Cell Factory HDPE Connectors, catalog # 171838, (2) White Tyvek Cover caps, catalog # 171897, (2) Blue sealing caps, catalog # 167652, a Gelman 4210 Bacterial Air Vent filter, approx 2 m of 8 mm I.D. silicone tubing and a tubing clamp), *see* **Note 5**.
2. PyrexPlus* aspirator bottle with tabulation, capacity 2 L (Corning, catalog # 61220-2 L) for loading Cell Factory.
3. 2 L Beaker for collection of Cell Factory waste.

2.3.3 Supplies for Closed System Only

1. Septum caps (Fisher) (Fisher, catalog #0292320).
2. Secondary Spike Set (Baxter) (VWR, catalog #68000-922).
3. Transfer Packs, 300 mL (Baxter, catalog #4R2014).
4. Transfer Packs, 600 mL (Baxter, catalog #4R2023).
5. Transfer Packs, 2,000 mL (Baxter, catalog #4R2041).
6. Bacterial air vent, sterilized (Pall Corporation, catalog #4308).
7. Plasma transfer set with 2 spikes (Baxter, catalog #4C2243).
8. TSCD Welding Wafers (Terumo, catalog #1SC*W017).
9. NS Adaptor with sterile dockable tubing (American Fluoroseal Corporation, catalog # 1Y24-18B)

2.4 Working Solutions

Complete Culture Medium (CCM): 500 mL α-MEM, 100 mL FBS (final conc.~16.5%), 6 mL L-Glutamine (final conc. 2 m*M*), and 6 mL Penicillin G and streptomycin sulfate (final conc. 100 units/mL penicillin and100 μg/mL streptomycin - *Optional).*

Filter medium through 0.22 μm sterile filter unit. Divide into aliquots and store at 4 °C for up to 2 wk. Before an experiment, warm the aliquot to 37 °C, *see* **Note 6**.

3 Methods

All supplies and reagents are sterile and the culture procedure should be performed in a tissue culture hood using aseptic techniques. All materials to be used in the BSC should be wiped down with 70% Ethanol before bringing it in to the BSC. Lab coats and gloves should be worn. Culture media and buffers should be pre-warmed to 37 °C before use.

3.1 Isolation of Human MSCs from Bone Marrow Aspirate

1. Under local anesthetic, bone marrow aspirates are collected from the iliac crest and placed in 10 mL Na Heparin Vacutainer tubes prefilled with 3 mL αMEM. The tubes and samples are kept on ice until transported to the laboratory and processed.
2. Each aspirate is transferred into a sterile 50 mL conical tube and diluted with sterile 1× Hank's Balanced Salt Solution (HBSS). QS for a total volume of 15 mL.
3. Each Vacutainer is then rinsed twice with an additional 5 mL of 1× HBSS and the contents combined with the diluted aspirate.
4. For each aspirate, place 10 mL of room temperature Ficoll into a separate 50 mL conical tube.
5. Gently overlay each aspirate onto the Ficoll, *see* **Note 7**.
6. Centrifuge at 1,800 *g* for 30 min at room temperature in a swinging bucket rotor with BRAKE OFF, *see* **Note 8**.
7. After centrifugation, collect the "buffy coat" located at the Ficoll-HBSS interface with a sterile Pasteur transfer pipet and place the cells into a clean 50 mL conical tube, *see* **Note 9**.
8. Dilute each sample with 1× Hank's Balanced Salt Solution (HBSS). QS for a total volume of 25 mL. Invert the tube 3–5 times to mix, *see* **Note 10**.
9. Centrifuge at 1,000 *g* for 10 min in a swinging bucket rotor with the BRAKE ON.
10. Remove the supernatant by vacuum aspiration and resuspend the cells with 30 mL of prewarmed CCM.
11. Plate the cells in a 15 cm^2 plate or T 175 cm^2 flask.
12. Incubate the cells at 37 °C with 5% humidified CO_2 for 18–24 hours to allow the adherent cells to attach.
13. Approximately 24 h later, remove the media and nonadherent cells, *see* **Note 11**.
14. Add 10 mL of prewarmed 1× PBS to the culture, rock gently to cover the entire surface area, and then remove the 1× PBS. Repeat the wash 2 additional times, *see* **Note 12**.

15. Add 30 mL of fresh CCM to the flask and return to the incubator. Incubate the cells at 37 °C with 5% humidified CO_2 for 5–10 d. Examine daily by phase microscopy. Every third day, remove the media, rinse the cells with 10 mL of prewarmed 1 × PBS, remove the PBS, and feed with a fresh 30 mL of CCM. Continue until the cells are between 60 and 80% confluent. Varying levels of confluence are shown in Fig. 1.1, *see* **Note 13**.
16. If the cells are going to be expanded in a Cell Factory, the Cell Factory must be equilibrated inside a humidified, 5% CO_2, 37 °C incubator for 48 h before use.
17. At harvest, remove the media and rinse the flask with 30 mL PBS. Remove the PBS. Add 10 mL of prewarmed Trypsin-EDTA solution to the flask. Distribute the trypsin across the surface area of the flask. Incubate the flask for 2–5 min at 37 °C. Examine the cells by phase microscopy.
18. After 80–90% of the cells have rounded up or become detached, gently tap the sides of the flask to dislodge any remaining attached cells, and add 10 mL CCM to the flask. Rock the flask back and forth to swirl the media around flask, and transfer the cell suspension into a clean 50 mL conical tube. Rinse the flask with 30 mL of 1 × PBS and combine with the cell suspension.
19. Centrifuge at 480 *g* for 10 min at room temperature in a swinging bucket rotor with the BRAKE ON.
20. Remove the supernatant and resuspend the cells with 1–2 mL of PBS or HBSS.
21. Count the cells with a hemocytometer and Trypan Blue or other method, *see* **Note 14**.

Viability of the cells can be checked using 0.4% Trypan Blue (Sigma). To 250 µL of Trypan blue add 150 µL HBSS and 100 µL cell suspension (dilution factor = 5). Wait 5 to 15 min. (If cells are exposed to Trypan blue for extended periods of time, viable cells as well as nonviable cells may begin to take up the dye). Count cells in 5 large squares on one side hemocytometer and calculate number of cells/mL using the formula:

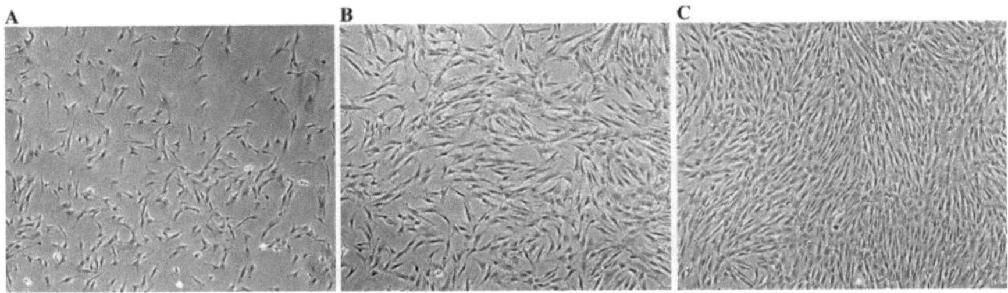

Fig. 1.1 Varying levels of confluence of MSCs. Figure 1.1A. This is approximately 40% confluent. It is too early to harvest this plate. Figure 1.1B. This is approximately 80% confluent. It is time to harvest this plate. Figure 1.1C. This is approximately 97% confluent. It is too late for optimal harvest of this plate. It is too confluent.

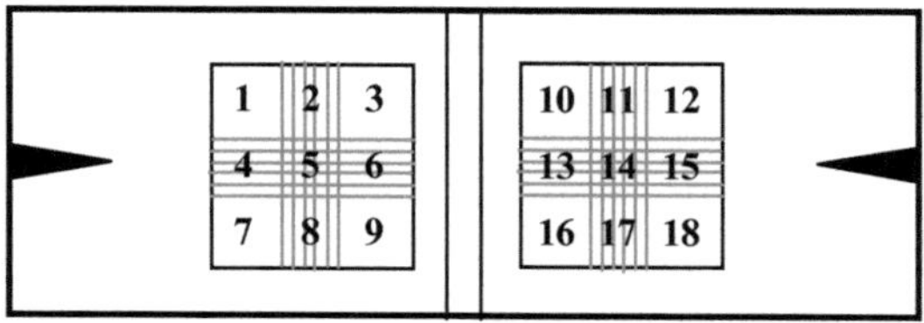

$$\frac{\textit{total cells counted X dilution factor X } 10^4}{\textit{number of large squares counted}} = \#\ cells/mL$$

The four corner squares and the center square (1,3,5,7,9 on one side or 10,12,14,16,18 on the other) in the figure above are considered the large squares.

Total number of cells: number of cells/mL X mLs

Nonviable cells will stain blue and live cells will be unstained. Count both blue and clear cells.

$$Viability\,(\%): \frac{\textit{Total viable cells (unstained)}}{\textit{Total cells (stained and unstained)}} \times 100$$

22. Seed cells at a density between 60 and 100 viable cells/cm^2 in the appropriate culture vessel. hMSCs can usually be successfully expanded through passage 4 and 5 without significant loss of stem cell phenotype.

The remainder of this procedure will describe large scale expansion of hMSCs in Nunc Cell Factories. Expansion in regular tissue culture flasks or dishes can be done, following the basic premise of plating at low density (60–100 cells/cm^2), feeding every 3–4d with fresh CCM and not allowing the cells to become more than 60–80% confluent.

3.2 Culture of hMSCs in Cell Factories using an Open or a Closed System

Cell Factories (Nunc) are used for large-scale cell culture and production of biomaterials such as vaccines, monoclonal antibodies and interferon. Cell Factories provide a large amount of growth surface in a small area with easy handling and low risk of contamination. They are designed for static cultures and can be used for anchorage dependent cells or cell suspensions. The Cell Factories are multiple trays, assembled by sonic welding, that are treated to produce a surface for good cell attachment. The chambers are connected by 2 port tubes for input and output of reagents, etc. The tubes have openings located in the upper half of each chamber. By having these openings high in the chamber, they are above media level during incubation. Turning the Cell Factory on its side puts one port tube at the lowest point of the unit. In this position the unit can be filled or drained.

There are 2 basic methods for loading, feeding and harvesting the Cell Factories. The less expensive, but more hands-on and cumbersome method is an Open System, which involves filling and draining the Cell Factory using a large sterile glass bottle containing the cell suspension, media, wash solution, trypsin or other reagents. The bottle has a bottom opening port that accepts tubing that then connects to the Cell Factory. Subsection 3.2.1 describes how to perform cell culture in Cell Factories using the Open System and Figs. 1.2 to 1.17 (printed with permission from Nalge Nunc International) illustrate the Cell Factory and the basic methodology of how it is set up, filled with reagent and drained using the Open System *(28)*.

The other technique for using Cell Factories for large-scale culture is a closed system. This is a more expensive technique because there is equipment that must be bought (a tube welder and tube sealer) and has a much higher cost of consumables (transfer bags, spikes, septum caps, and welding wafers). However, this method is much more convenient in terms of handling the reagents, loading and

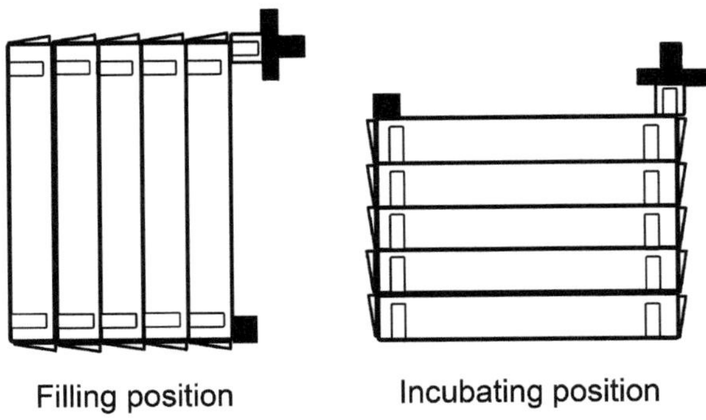

Fig. 1.2 Illustration of the Filling and Incubating Positions for the Cell Factory.

Fig. 1.3 Unpack the Cell Factory and place in a laminar air flow cabinet to work under sterile conditions.

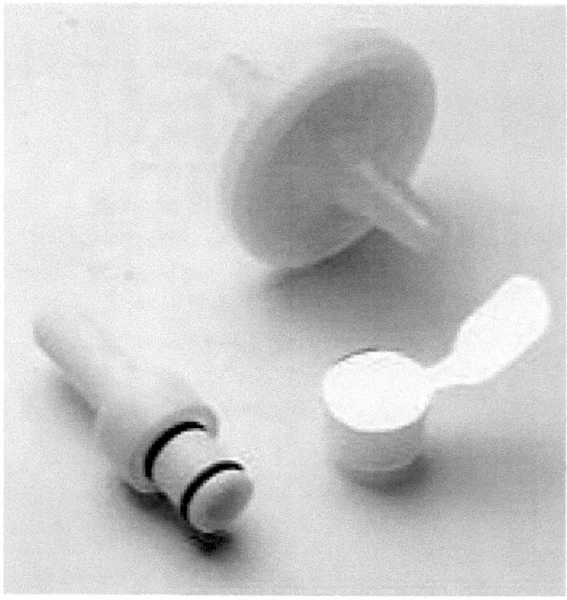

Fig. 1.4 Accessories: Air filter, tube connector, adaptor cap.

Fig. 1.5 Remove the seal from one of the adaptor caps.

Fig. 1.6 Immediately insert the pre-sterilized air filter in the cap.

Fig. 1.7 Aspirator bottle with cell suspension, mounted with sterile tubing, tube connector and clamp.

Fig. 1.8 Remove the second adaptor cap from the Cell Factory.

Fig. 1.9 Insert the tube connector.

Fig. 1.10 Turn Cell Factory on its side. Raise aspirator bottle above Cell Factory level.

Fig. 1.11 Agitate the aspirator bottle, loosen the clamp and the cell suspension will flow into the Cell Factory. The chambers will initially fill unevenly owing to a difference in pressure.

Fig. 1.12 When filling is completed, the levels of liquid in all chambers will equalize through the connecting channel.

Fig. 1.13 Turn the Cell Factory 90°, putting the filling inlet up. The medium will be separated, with equal amounts of medium in each chamber.

Fig. 1.14 Place the Cell Factory in a horizontal position.

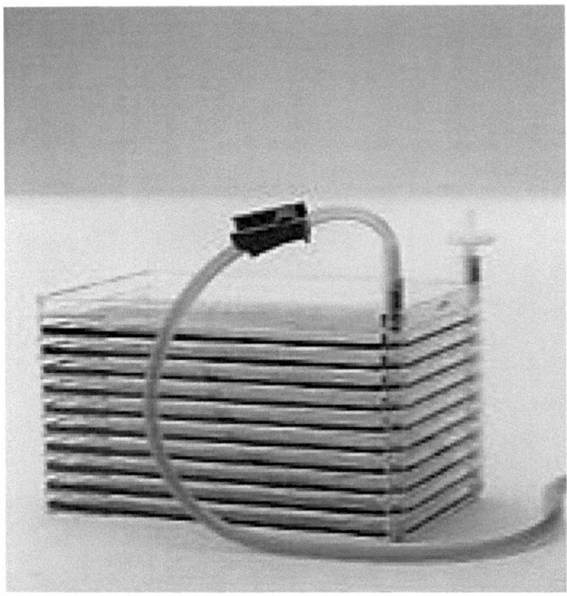

Fig. 1.15 The surface of all the trays will be covered by medium. Remove the tube connector, mount the adaptor cap, but leave the filter on, and incubate at 37 °C with 5% CO2. Ensure that the Cell Factory is sitting level in the incubator, so that media is at an equal level in all trays.

Fig. 1.16 To empty the Cell Factory, place it above the aspirator bottle, in the position shown, and the liquid will run out. Alternatively, the liquid can be poured out of that port into a waste beaker.

Fig. 1.17 Tilt the Cell Factory to drain completely.

draining the Cell Factory, removing the waste, and in keeping possibility of contamination at a minimum. Although this is important in in vitro experimentation, it is of particular importance for MSCs that may be used for in vivo studies to diminish the possibility of infection. Section 3.2.2 describes how to perform cell culture using the Closed System handling of Cell Factories.

3.2.1 Procedure for Open System Handling of Cell Factory

1. Before seeding Cell Factory, be sure 2L glass aspirator bottle with tubing, connector and clamp have been autoclaved and are sterile. Be sure tubing end with connector and bottle mouth are covered with foil when autoclaving.
2. Prepare 1.2L CCM using double the volumes given in Section 2.4, Working Solutions. Filter using 1L Stericup. Place in 37 °C water bath until use, *see* **Note 15**.
3. From harvested cell suspension, withdraw the appropriate volume to give 300,000 (~60 cells/cm^2) or 600,000 (~100 cells/cm^2) MSCs with a pipetor (i.e., 100–1000 μL) and dispense the cells directly into pre-warmed 1.2L CCM. Replace cap and mix by gently inverting, taking care to avoid bubbles.
4. If an inverted phase microscope with a SLWDC is available, please skip to Step 6.
5. If there is no phase microscope with a SLWDC available, a "control" plate must be made to monitor the approximate cell growth in the Cell Factory. Remove 10 mL or 15 mL of cell suspension from 1.2L of cell suspension with a sterile pipet and add this to a 10 cm diameter dish or T75 flask, respectively, as a control plate. This control plate is used to monitor the MSC growth microscopically to estimate MSC growth in the Cell Factory. Incubate at 37 °C with 5% CO_2, *see* **Note 16**.
6. Add the media containing the MSCs to the sterile glass aspirator bottle fitted with the tubing, tube connector and clamp. Mix thoroughly by swirling.
7. To load the Cell Factory, lay the Cell Factory on its side as shown in Fig. 1.10 and place the aspirator bottle of CCM containing the cells above the Cell Factory. Open the clamp and let the cells drain into the Cell Factory, as described in Fig. 1.11. Rotate the Cell Factory to horizontal and rock it gently side-to-side and back-to-front to evenly cover all levels of the Cell Factory.
8. Incubate the Cell Factory (and control plate) at 37 °C with 5% humidified CO_2 for 5–10 d. Examine the Cell Factory (or the control plate) daily by phase microscopy.
9. Every third day, remove the media from Cell Factory by draining it and feed the Cell Factory with 1.2L of fresh, prewarmed CCM, following Cell Factory draining and filling instructions outlined above. A large beaker can be used to collect the waste. Rotate Cell Factory to horizontal and rock it gently side-to-side and back-to-front to evenly wash all levels of the Cell Factory, *see* **Note 17**.
10. Continue incubation of Cell Factory (and control plate) culture until the cells are between 60 and 80% confluent.

11. When cells have reached 60–80% confluence, either determined directly from examining the Cell Factory itself or by using the "control" plate as an estimate, harvest the Cell Factory (and the control plate).
12. Be sure to have the aspirator bottle, tubing, tubing connector and clamp sterilized as in Step 1.
13. To harvest the Cell Factory, drain culture media from Cell Factory into a 2L waste beaker. (To harvest the control plate, remove culture media from dish or flask by aspiration or by pipeting and discarding into 2L waste beaker).
14. Add 200mL warmed PBS to Cell Factory using the sterile aspirator bottle with tubing and connector. (For the "control" plate, add 3mL PBS to a 10cm diameter dish or add 5mL PBS to a T75 flask, using a pipet), *see* **Note 18**.
15. Rotate Cell Factory to horizontal and rock it gently side-to-side and back-to-front to evenly wash all levels of the Cell Factory.
16. Drain PBS from the Cell Factory into the 2L waste beaker (For "control" plate, remove PBS by aspiration).
17. Add 200mL of prewarmed 0.25% trypsin in 1mM EDTA to Cell Factory using aspirator bottle and tubing. (For "control" plate, add 3mL trypsin/EDTA to 10cm diameter dish or 5mL trypsin/EDTA to T75 flask). Place Cell Factory (and control plate) in 37°C CO_2 incubator for 2–5min. Monitor cell detachment microscopically either on the Cell Factory itself or by using the "control" plate as a monitor. When about 90% of the cells are detached, stop the reaction.
18. Add 200mL prewarmed CCM to the Cell Factory using the aspirator bottle and tubing. Rotate the Cell Factory to horizontal and rock it gently side-to-side and back-to-front to evenly cover all levels of the Cell Factory. (For "control" plate, add 3mL prewarmed CCM to 10cm diameter dish or 5mL prewarmed CCM to T75 flask), *see* **Note 19**.
19. Drain the contents of Cell Factory (cells in CCM-neutralized trypsin) into a sterile container with a 1L capacity. (Transfer the contents of the "control" plate to a 15mL plastic sterile conical tube).
20. Add 200mL PBS to Cell Factory using aspirator bottle and tubing. Rotate the Cell Factory to horizontal and rock it gently side-to-side and back-to-front to evenly cover all levels of the Cell Factory. (For "control" plate, add 3mL PBS to a 10cm diameter dish or add 5mL PBS to a T75 flask).
21. Drain PBS into the 1L sterile container with the cell suspension. (For "control" plate, add PBS wash to 15mL conical tube containing harvested cells).
22. Distribute the cell suspension equally into four 200mL sterile plastic conical tubes.
23. Centrifuge at 480*g* for 10min at room temperature in a swinging bucket rotor with the BRAKE ON. (For "control" plate, centrifuge the 15mL conical tube containing the cell suspension with a balance tube at these parameters).
24. Open the conical tubes and remove and discard the supernatant. Resuspend each pellet in 2mL of CCM or HBSS. Resuspend the pellet by drawing the cell suspension gently into and out of the pipet 10–15 times. Repeat for the remaining tubes. (For "control" plate, remove supernatant from 15mL conical tube. Add 1mL CCM or HBSS and resuspend cell pellet by gently pipeting with a 1,000μL pipetor), *see* **Note 20**.

25. Combine the contents of the four 200 mL tubes into one. Rinse each tube with 25 mL of CCM or HBSS and combine with the cell suspension.
26. Centrifuge at 480 *g* for 10 minutes at room temperature in a swinging bucket rotor with the BRAKE ON.

If a centrifuge with carriers that can hold the 200 mL tubes is not available, an alternative method is to distribute the cell suspension over 50 mL conical tubes, starting with Step 22.

22a. Pipet 50 mL aliquots of mixed cell suspension into twelve (12) 50 mL conical centrifuge tubes.
23a. Centrifuge (Brake ON) at 480 *g*, RT for 10 minutes.
24a. Aspirate supernatant. Add 1 mL CCM or HBBS to each tube and resuspend cells. Combine all cell suspensions into one 50 mL conical tube.
25a. Wash each tube with 2 mL CCM or HBBS and add the washes to the 50 mL conical tube containing the cells.
26a. Centrifuge (Brake ON) at 480 *g*, RT for 10 min.

The procedure now continues from Step 26 above.

27. Remove the supernatant. Resuspend the cells in a small volume (typically 2–5 mL) of CCM or HBSS. Count the cells with a hemocytometer and Trypan Blue or other method, *see* **Note 21**.
28. Use the cells for experimental purposes, reseed additional flasks or factories at a density of 60–100 cells/cm^2, or cryopreserve the cells (see Chapter 8 "Freezing harvested hMSCs and Recovery of hMSCs from Frozen Vials for subsequent expansion, analysis and experimentation").
29. Assess hMSC characteristics and quality by: Colony Forming Units (CFU), FACs analysis of selected cell surface markers, and the ability of hMSCs to differentiate to osteoblasts, adipocytes, and chondroblasts in vitro.

3.2.2 Procedure for Closed System Handling of Cell Factory

1. Withdraw the appropriate volume of cell suspension with a pipetor (i.e. 100–1000 μL) and dispense the cells directly into a prewarmed bottle of CCM (1.2L). Replace the manufacturers cap with a sterile septum cap. Mix by swirling the bottle.
2. Attach an NS adaptor with sterile dockable tubing to adaptor cap on inlet port of Cell Factory. Pierce the septum cap with the spike and tube assembly found in the secondary spike set.
3. Weld the tube assembly from the Cell Factory to the tube assembly of the bottle of CCM containing the cells with the Terumo™ TSCD Sterile Tubing Welder (model SC-201A).
4. Hold the bottle of CCM containing the cells above the Cell Factory. Open the blue pinch clamp, break the seal from the sterile weld, and let the cells drain into the Cell Factory. Hand seal the tube assembly with the Sebra™ hand sealer model 2380, and discard the tubing assembly.

Activity	Size of transfer bag	Reagent volume	Material/reagent	Size transfer bag for collection of contents
Feeding	2,000 mL	1,200 mL	CCM	2,000 mL
Harvest	300 mL	200 mL	Trypsin	2,000 mL
Harvest	300 mL	200 mL	PBS	2,000 mL
Harvest	300 mL	200 mL	CCM	2,000 mL

5. Level the Cell Factory to equally distribute the cells throughout all 10 layers. Rock the Cell Factory back & forth to ensure the cell suspension covers the entire surface area on each layer.
6. Incubate the cells at 37 °C with 5% humidified CO_2 for 5–10 d. Examine daily by phase microscopy. Every third day, remove the media and feed with a fresh 1.2 L of CCM. Continue until the cells are between 60 and 80% confluent.
7. One method for culturing cells in Cell Factories is to prefill sterile transfer bags with needed cell culture reagents (i.e., trypsin, 1× PBS, 1× HBSS, and CCM) and use empty transfer bags to collect the contents of the Cell Factory. The table above lists typical culture activities, the type of transfer bag required, and volume of reagent used. Reagents are placed into transfer bags by replacing the manufacturers cap with a sterile septum, piercing the septum cap with the spike and tube assembly, and then sterile welding the tube assembly to the transfer bags tubing. After transfer, the tubing is sealed with the hand sealer and discarded.

For feeding or harvest, pre-warm all reagents to 37 °C.

8. For feeding, remove Cell Factory from incubator. Sterile weld a 2,000 mL transfer pack to the Cell Factory. Place the Cell Factory on its side with the transfer pack at a lower height than the Cell Factory. Break the seal and allow the medium to drain into the bag. Seal the tube with the hand held tube sealer, remove and discard the bag.
9. Sterile weld a warmed transfer bag with 1.2 L CCM to the Cell Factory. Hold the bag higher than the Cell Factory, break the seal, and allow the CCM to drain into the Cell Factory. Distribute CCM equally amongst all layers. Seal the tube to the empty transfer bag with the hand sealer, remove and discard the bag. Return Cell Factory to incubator.
10. For harvest, remove Cell Factory from incubator, sterile weld a 2,000 mL transfer pack to the Cell Factory. Place the Cell Factory on its side with the transfer pack at a lower height than the Cell Factory. Break the seal and allow the medium to drain into the bag. Seal the tube with the hand held tube sealer, remove and discard the bag.
11. Sterile weld a warmed transfer bag with 200 mL PBS to the Cell Factory. Hold the bag higher than the Cell Factory, break the seal, and allow the PBS to drain into the Cell Factory. Distribute PBS equally amongst all layers. Drain the PBS rinse back into the attached bag. Seal the tube with the hand sealer, remove and discard the bag.

12. Sterile weld to one side of the Y-tubing set a 300 mL transfer bag containing 200 mL prewarmed trypsin. To the other side, sterile weld a 2 L transfer bag containing 200 mL of CCM. Make sure the clamps are closed.
13. Hold the bag containing the trypsin higher than the Cell Factory, break the seal, open the clamp, and allow the trypsin to drain into the Cell Factory. Seal the tubing with the hand held tube sealer. Remove and discard the spent transfer bag. Equally distribute the trypsin amongst the layers. Incubate the Cell Factory for 2 to 5 minutes at 37 ± 1 °C. Verify by phase microscopy that 80–90% of the cells have rounded up or become detached. Tap the sides of the Cell Factory to dislodge any remaining attached cells.
14. Place the Cell Factory on its side so that the tubing set is in the lower corner and the filter is in the upper corner. Hold the bag containing the CCM higher than the Cell Factory. Break the seal, open the clamp, and allow the CCM to drain into the Cell Factory. Equally distribute the CCM amongst the layers. Drain the contents of the Cell Factory back into the 2 L transfer bag. Close the roller clamp to the partially filled 2 L transfer bag and leave attached to Cell Factory.
15. Sterile weld a warmed transfer bag with 200 mL PBS to the other side of the Y-tubing set. Hold the bag higher than the Cell Factory, break the seal, and allow the PBS to drain into the Cell Factory. Seal the tubing with the hand held tube sealer, remove and discard the empty PBS bag. Distribute PBS equally amongst all layers. Drain the PBS rinse into the 2 L transfer bag containing the trypsinized cells.
16. Continue cell processing using Steps 22–29 from "Procedure for Open System of Handling Cell Factories".

4 Notes

1. FBS is lot selected by culturing hMSCs in CCM using several different lots of FBS and comparing MSC growth, expansion, morphology, CFU and differentiation potential. FBS is a critical component for these cells and it is imperative that a premium grade of FBS be used.
2. To view cultures in Cell Factories that have more than 2 trays, it is necessary that the microscope be fitted with Super Long Working Distance Condenser. The Nikon Eclipse TE2000 has this accessory. If a microscope with SLWDC is not available, a control plate using a regular T75 flask should be seeded at the same time as the Cell Factory(ies) and used to approximate the rate of cell growth in the Cell Factory. There are also 1-tray and 2-tray Cell Factories available that can be monitored using an inverted phase microscope with regular condenser length, but the surface area available for culture is much less.
3. If no microscope with SLWDC is available, regular tissue culture dishes or flasks will be needed for control plates for Cell Factory. Those with the same type of plastic as the Cell Factories are: Sterile tissue culture dishes/flasks: 10 cm diameter (~57 cm^2) dishes (Nunc; Catalog # 150350), 15 cm diameter (145 cm^2) dishes (Nunc; Catalog # 168381) or T75 (75 cm^2) flasks (Nunc; Catalog # 156499). An additional option is to use 1- or 2-tray Cell Factories rather than 10 tray Cell Factories: 1-tray Cell Factory, 632 cm^2 culture area (Nunc, Catalog # 165250); 2-tray Cell Factory, 1,264 cm^2 culture area (Nunc, Catalog # 167695).
4. These units are filter safe for DMSO, which is required when filtering freezing media.

5. Additional accessory and replacement parts are available from Nunc. CellStack® multilayer culture chambers from Corning are probably of equal quality, but we have not tested them as extensively. However, early testing showed comparable results to Nunc Cell Factories.
6. It is preferable to grow cells without penicillin/streptomycin because any contamination of the culture will not be masked by the presence of antibiotics. It is advisable not to use antibiotics if cells are to be used for *in vivo* experiments.
7. If the Ficoll and HBSS-cell suspension layers are admixed, the mononuclear cells will not completely separate out during centrifugation.
8. The brake is left off to ensure even deceleration of the centrifuge and thereby lessen the possibility of disturbing Ficoll-HBSS cell suspension interface.
9. Remove the cells in as small a volume as possible. This enhances the removal of Ficoll by dilution.
10. The volume ratios of diluent to sample should be at least 3:1.
11. If the nonadherent cells are not removed, hematopoietic cells may become attached and contaminate the hMSC culture.
12. There may not be very many adherent cells seen at this point.
13. To preserve stem cell phenotype, do not allow the cells to become confluent. Because hMSCs are not evenly distributed in the marrow, some aspirates do not have enough hMSCs to obtain large cultures. If a sample does not grow well or have good morphology by the 8th day, discard it.
14. Typical yields from a T-175 flask are between 1 and 3×10^6 total cells, with an average viability usually greater than 90%.
15. The 1 L Stericup is large enough to hold 1.2 L CCM.
16. The control plate should have the same type of growing surface (Nunclon™Δ) as the Cell Factory to have similar conditions for cell growth and thus give a more accurate monitor for cell growth in Cell Factory.
17. To feed the control plate, remove spent media using aspiration or a sterile pipet. (Can use waste beaker mentioned above). Using a pipet, add 10 mL fresh, pre-warmed CCM to a 10 cm diameter dish or add 15 mL fresh, pre-warmed CCM to a T75 flask.
18. This step is necessary to remove any residual CCM with FBS, which can inhibit the action of the trypsin.
19. Complete culture media containing FBS must be used so that the trypsin will be neutralized and it is important to distribute the CCM over all levels.
20. The cells from the "control" plate can now be counted using Trypan Blue and a hemocytometer. The count can then be compared to the count obtained from the Cell Factory harvest to see if the growth estimate was accurate.
21. Typical yields from a Cell Factory are between 30 and 90×10^6 total cells, with an average viability usually greater than 90%.
22. Troubleshooting - Different preparations of hMSCs vary because of variations in the number and quality of cells obtained with blind aspirates of bone marrow. It may be necessary to determine the optimal conditions for growth for each laboratory; using different FBS, trying different plating densities and/or growing the cells for different time intervals. Several factors to consider for troubleshooting the hMSC cultures are shown below:

    a. Frequently monitor the cultures by phase microscopy. With practice you will be able to identify cultures that have high RS content (long spindly cells, not highly vacuolated) and expand rapidly.
    b. Use a good quality lot of FBS. This is critical in hMSC cell culture. Different lots of FBS must be tested before use in culture to determine the FBS lot that gives the best overall cell growth and functionality. Test 3–5 lots and reserve with the supplier the lot that gives the best rate of growth, highest CFU activity, best differentiation and best morphology.
    c. The growth rate and content of RS cells in the cultures is a function of passage number, plating density, and time of incubation. If cells are not growing well or lose their spindle-like morphology, be sure you are working with early passage cells, that they are plated at low density and not allowed to grow to confluence before harvest.

d. Monitor color and level of media over time to make certain pH is correct. The color should remain pink in color. If the media turns yellow (acid) or purple (base), this is usually an indication of bacteria, fungal or other type of contamination. Make sure the plates are level in the incubator and do not allow evaporation to substantially reduce level of media in dish.
e. If fungal or bacterial contamination develops, discard the culture and check incubator for signs of contamination. If contamination is seen, the incubator and the culture laboratory must be decontaminated.
f. One of the most common errors is not ensuring that the cells are adequately dispersed in solutions before counting or plating. They settle quickly.

References

1. Liechty, K. W., MacKenzie, T. C., Shaaban, A. F., Radu, A., Moseley, A. M., Deans, R., Marshak, D. R., and Flake, A. W. (2000) Human mesenchymal stem cells engraft and demonstrate site-specific differentiation after in utero transplantation in sheep. *Nat. Med.* **6**, 1282–1286.
2. Pereira, R. F., Halford, K. W., O'Hara, M. D., Leeper, D. B., Sokolov, B. P., Pollard, M. D., Bagasra, O., and Prockop, D. J. (1995) Cultured adherent cells from marrow can serve as long-lasting precursor cells for bone, cartilage, and lung in irradiated mice. *Proc. Natl. Acad. Sci. U.S.A.* **92**, 4857–4861.
3. Kopen, G. C., Prockop, D. J., and Phinney, D. G. (1999) Marrow stromal cells migrate throughout forebrain and cerebellum, and they differentiate into astrocytes after injection into neonatal mouse brains. *Proc. Natl. Acad. Sci. U.S.A.* **96**, 10,711–10,716.
4. Azizi, S. A., Stokes, D., Augelli, B. J., Digirolamo, C., and Prockop, D. J. (1998) Engraftment and migration of human bone marrow stromal cells implanted in the brains of albino rats–similarities to astrocyte grafts. *Proc. Natl. Acad. Sci. U.S.A.* **95**, 3908–3913.
5. Chopp, M., Zhang, X. H., Li, Y., Wang, L., Chen, J., Lu, D., Lu, M., and Rosenblum, M. (2000) Spinal cord injury in rat: treatment with bone marrow stromal cell transplantation. *Neuroreport* **11**, 3001–3005.
6. Friedenstein, A. J., Petrakova, K. V., Kurolesova, A. I., and Frolova, G. P. (1968) Heterotopic of bone marrow.Analysis of precursor cells for osteogenic and hematopoietic tissues. *Transplantation* **6**, 230–247.
7. Owen, M. and Friedenstein, A. J. (1988) Stromal stem cells: marrow-derived osteogenic precursors. *Ciba Found. Symp.* **136**, 42–60.
8. Caplan, A. I. (1990) Stem cell delivery vehicle. *Biomaterials* **11**, 44–46.
9. Prockop, D. J. (1997) Marrow stromal cells as stem cells for nonhematopoietic tissues. *Science* **276**, 71–74.
10. Krause, D. S., Theise, N. D., Collector, M. I., Henegariu, O., Hwang, S., Gardner, R., Neutzel, S., and Sharkis, S. J. (2001) Multi-organ, multi-lineage engraftment by a single bone marrow-derived stem cell. *Cell* **105**, 369–377.
11. LaBarge, M. A. and Blau, H. M. (2002) Biological progression from adult bone marrow to mononucleate muscle stem cell to multinucleate muscle fiber in response to injury. *Cell* **111**, 589–601.
12. Pittenger, M. F., Mackay, A. M., Beck, S. C., Jaiswal, R. K., Douglas, R., Mosca, J. D., Moorman, M. A., Simonetti, D. W., Craig, S., and Marshak, D. R. (1999) Multilineage potential of adult human mesenchymal stem cells. *Science* **284**, 143–147.
13. Colter, D. C., Sekiya, I., and Prockop, D. J. (2001) Identification of a subpopulation of rapidly self-renewing and multipotential adult stem cells in colonies of human marrow stromal cells. *Proc. Natl. Acad. Sci. U.S.A.* **98**, 7841–7845.

14. Sekiya, I., Larson, B. L., Smith, J. R., Pochampally, R., Cui, J. G., and Prockop, D. J. (2002) Expansion of human adult stem cells from bone marrow stroma: conditions that maximize the yields of early progenitors and evaluate their quality. *Stem Cells* **20**, 530–541.
15. DiGirolamo, C. M., Stokes, D., Colter, D., Phinney, D. G., Class, R., and Prockop, D. J. (1999) Propagation and senescence of human marrow stromal cells in culture: a simple colony-forming assay identifies samples with the greatest potential to propagate and differentiate. *Br. J. Haematol.* **107**, 275–281.
16. Colter, D. C., Class, R., DiGirolamo, C. M., and Prockop, D. J. (2000) Rapid expansion of recycling stem cells in cultures of plastic-adherent cells from human bone marrow. *Proc. Natl. Acad. Sci. U.S.A.* **97**, 3213–3218.
17. Prockop, D. J., Sekiya, I., and Colter, D. C. (2001) Isolation and characterization of rapidly self-renewing stem cells from cultures of human marrow stromal cells. *Cytotherapy* **3**, 393–396.
18. Sanchez-Ramos, J., Song, S., Cardozo-Pelaez, F., Hazzi, C., Stedeford, T., Willing, A., Freeman, T. B., Saporta, S., Janssen, W., Patel, N., Cooper, D. R., and Sanberg, P. R. (2000) Adult bone marrow stromal cells differentiate into neural cells *in vitro*. *Exp. Neurol.* **164**, 247–256.
19. Woodbury, D., Schwarz, E. J., Prockop, D. J., and Black, I. B. (2000) Adult rat and human bone marrow stromal cells differentiate into neurons. *J. Neurosci. Res.* **61**, 364–370.
20. Kotton, D. N., Ma, B. Y., Cardoso, W. V., Sanderson, E. A., Summer, R. S., Williams, M. C., and Fine, A. (2001) Bone marrow-derived cells as progenitors of lung alveolar epithelium. *Development* **128**, 5181–5188.
21. Wakitani, S., Saito, T., and Caplan, A. I. (1995) Myogenic cells derived from rat bone marrow mesenchymal stem cells exposed to 5-azacytidine. *Muscle Nerve* **18**, 1417–1426.
22. Fukuda, K. (2002a) Molecular characterization of regenerated cardiomyocytes derived from adult mesenchymal stem cells. *Congenit. Anom. Kyoto* **42**, 1–9.
23. Fukuda, K. (2002b) Reprogramming of bone marrow mesenchymal stem cells into cardiomyocytes. *C. R. Biol.* **325**, 1027–1038.
24. Fukuda, K. (2003) Use of adult marrow mesenchymal stem cells for regeneration of cardiomyocytes. *Bone Marrow Transplant.* **32** (Suppl 1), S25–S27.
25. Barbash, I. M., Chouraqui, P., Baron, J., Feinberg, M. S., Etzion, S., Tessone, A., Miller, L., Guetta, E., Zipori, D., Kedes, L. H., Kloner, R. A., and Leor, J. (2003) Systemic delivery of bone marrow-derived mesenchymal stem cells to the infarcted myocardium: feasibility, cell migration, and body distribution. *Circulation* **108**, 863–868.
26. Gronthos, S., Zannettino, A. C., Hay, S. J., Shi, S., Graves, S. E., Kortesidis, A., and Simmons, P. J. (2003) Molecular and cellular characterisation of highly purified stromal stem cells derived from human bone marrow. *J. Cell Sci.* **116**, 1827–1835.
27. Barry, F., Boynton, R., Murphy, M., Haynesworth, S., and Zaia, J. (2001) The SH-3 and SH-4 antibodies recognize distinct epitopes on CD73 from human mesenchymal stem cells. *Biochem. Biophys. Res. Commun.* **289**, 519–524.
28. Nunc Cell Factory Instruction brochure. (2004) Instructions and photos used with permission of Nalge Nunc International. Website address: http://www.nuncbrand.com/page/GB/10781.aspx, Nalge Nunc International USA.

Chapter 2
Mesenchymal Stem Cells from Adult Bone Marrow

Mark F. Pittenger

Abstract Mesenchymal stem cells (MSCs), sometimes referred to as marrow stromal cells or multipotential stromal cells, represent a class of adult progenitor cells capable of differentiation to several mesenchymal lineages. They can be isolated from many tissues although bone marrow has been used most often. The MSCs may prove useful for repair and regeneration of a variety of mesenchymal tissues such as bone, cartilage, muscle, marrow stroma, and the cells produce useful growth factors and cytokines that may help repair additional tissues. There is also evidence for their differentiation to nonmesenchymal lineages, but that work will not be considered here. This chapter will provide the researcher with some background, and then provide details on MSC isolation, expansion and multilineage differentiation. These are the beginning steps toward formulating tissue repair strategies. The methods provided here have been used in many laboratories around the world and the reader can begin by following the methods presented here, and then test other methods if these prove unsatisfactory for your intended purpose.

Keywords Mesenchymal stem cells (MSCs); direct plating isolation; density gradient isolation, lineage differentiation protocols; chondrogenic; adipogenic; osteogenic.

1 Introduction

Tissue healing takes place more rapidly in children than in adults and this is likely owing to a number of factors. One of those factors appears to be the abundance of stem and progenitor cells in the developing tissues of the child. As we reach adulthood, these cells are not so necessary for tissue growth and appear to diminish over time, perhaps some differentiating to adult cell types whereas some are retained as resident tissue stem cells. Over the years, the number of tissue resident stem cells further diminishes as they are called on for normal tissue repair and maintenance,

From: *Methods in Molecular Biology, vol. 449, Mesenchymal Stem Cells: Methods and Protocols*
Edited by D.J. Prockop, D.G. Phinney, and B.A. Bunnell © Humana Press, Totowa, NJ

and normal cellular senescence. That said, there appear to be small numbers of progenitor or stem cells that can be isolated from many tissues at all stages of life. These cells appear to afford a wonderful opportunity, indeed, a responsibility, to understand important aspects of human biology involving tissue repair and regeneration.

One of these adult stem cells that can be found in several tissues throughout life and that can be isolated and propagated in culture was termed the mesenchymal stem cell or MSC *(1,2)* by Arnold Caplan of Case Western University. A key element in the acceptance of MSCs as a potential cellular therapeutic was the early demonstration of safety in humans by Drs. Hillard Lazarus and Stan Gerson, hematological oncologists, who first tested MSCs as support cells during hematopoietic stem cell (HSC) transplantations *(3)*. Work began on the isolation of MSCs and examination of their multipotential nature in 1994, and the multilineage in vitro differentiation of these cells was demonstrated at the 1996 annual meeting of the American Society for Cell Biology. This work led to a key paper in the stem cell field published in Science in 1999 demonstrating multilineage differentiation of clonal populations of human MSCs that has now been cited in over 2,500 publications *(4)*.

The first descriptions of fibroblastic cells that could be isolated and grown from bone marrow samples, that retained the ability to differentiate to bone tissue was presented by Dr. Alexander Friedenstein of the Gamalaya Institute in Moscow in the 1960s, using guinea pig bone marrow as the source *(5–7)*. When these cells were culture expanded ex vivo, and then placed in capsules under the skin of a recipient syngeneic animal, new bone and cartilage tissue was identified when histology was performed. Although the same type of cell, or a close homologue of it, can be found in many tissues, including adipose *(8–11)*, the endosteal surface of bone and bone itself, bone marrow has proven to be a reproducible and convenient source of these cells from all species tested. MSCs have been isolated from mouse *(12–16)*, rat *(17–21)*, guinea pig *(5,6)* rabbit *(22–24)*, dog *(25,26)*, goat *(27)*, pig *(28–31)*, nonhuman primates *(32–35)*, and man *(1–4,36–40)*.

MSCs secrete growth factors and cytokines that have autocrine and paracrine activities. The MSCs produce vascular endothelial growth factor (VEGF), stem cell factor (SCF-1), leukemia inhibitory factor (LIF), granulocyte colony stimulatory factor (G-CSF), macrophage colony stimulating factor (M-CSF), granulocyte-macrophage colony stimulating factor(GM-CSF), interleukins (IL-1, -6,-7, -8, -11, -14, and -15), stromal cell-derived factor (SDF-1), Flt-3 ligand, and others *(4,36–39)*. The expression of these factors may be modulated through interactions with other cell types *(48–50)*.

Some additional interesting and important aspects of MSCs that have come to light include their homing to sites of tissue injury, particularly ischemic regions of heart *(26,29,43,44)* where the MSCs may prevent deleterious remodeling *(28,30,31)*. MSCs also have the ability to modify immune responses and engraft in allogeneic recipients, and MSC treatment has been used to clinically treat graft-versus-host disease (GVHD) *(46–51)*. MSCs are also under evaluation for clinical use in children with osteogenesis imperfecta, and glycogen storage diseases *(52–54)*. Although methods in these areas are not detailed here, clearly, MSCs represent a new, exciting and potentially powerful paradigm for cellular therapy.

Although a number of research groups investigated MSCs from nonhuman species, Arnold Caplan, Steve Haynesworth, and colleagues at Case Western Reserve

University were the first to systematically pursue the isolation of the MSC from human tissue. The reasoning was that a human source should be sought that could be harvested as a simple procedure in the doctors office under local anesthetic, without sacrificing or harming the tissue that was to be repaired and regenerated. In this case, bone marrow was chosen as it was remarkably renewable, harbored MSCs, was a known source of hematopoietic stem cells (HSCs), and could be isolated from the marrow cavity of the hipbone in a simple procedure under local anesthetic. I can personally attest to the simplicity of the harvest procedure. This will be the method used to obtain the starting material below. The methods presented here work well for most species, but the mouse MSCs are often contaminated with HSCs that seem to be required during propagation of MSCs, perhaps because of production of a growth factor or cytokine.

As a senior scientist at Osiris Therapeutics, Inc. a company formed to develop and commercialize products based on human MSCs, I was able to perform studies on MSCs and evaluate many of their properties. The results have been presented in a number of peer-reviewed papers that form the basis for this chapter and all the information here has been made available previously. The first public presentation of the 3 lineage differentiation of MSCs to chondrocytes, adipocytes, and osteocytes was at the American Society for Cell Biology Annual Meeting in December 1996 in Washington, D.C.

2 Materials

2.1 MSCs from Bone Marrow Aspirates

2.1.1 Biohazard Considerations

Bone marrow may contain blood borne pathogens, and therefore one must take care to avoid exposure, splashing, or spills. The handler should be trained and familiar with "universal precautions" to protect exposed skin, mucus membranes etc, and it is a good idea to walk through the planned steps before obtaining the bone marrow sample. Recommendations can be found in the Code of Federal Regulations, 29 CFR Chapter XVII (7-1-98 Edition), Occupational Safety and Health Administration, US Department of Labor. § 1910.1030. Although the bone marrow donors may be pretested for HIV and hepatitis, such tests are not available for many pathological agents.

2.2 Cell Culture Laboratory

It is understood that culturing MSCs will require typical equipment found in a cell culture laboratory including a laminar flow Biological Safety Cabinet, a 5% CO_2 incubator at 37 °C, inverted microscope with interference phase optics for observing cultured cells, and a hemocytometer (glass) for counting cells with the microscope.

2.3 Bone Marrow Acquisition

Human bone marrow can be obtained by needle from the iliac crest by a physician in a brief procedure, using local anesthetic. For many laboratories, this means finding a medical doctor, a hematologist/oncologist, who is willing to collaborate. For harvest of bone marrow from nonhuman species the procedure is similar but the subject is appropriately anesthetized. For rodents, they are humanely sacrificed by IACUC approved procedures. In general, for the iliac crest biopsy, the skin area is carefully cleaned, lidocaine is injected locally to numb the area, and a fine trocar is used to gain access to the marrow space. The aspiration syringe is loaded with 3,000 units of heparin to prevent clotting of the marrow sample. A large gauge needle is inserted and marrow is aspirated with rapid pulls on the syringe plunger, moving the needle to gain access to new marrow areas without removing the needle. Slow steady pulling on the plunger is not desirable as this will likely aspirate less marrow, and more blood as it flows into the space. Successful marrow aspiration is usually limited to 15–20 mLs per side, left and right iliac crest. More volume usually yields more blood, not marrow. This step is perhaps the most important in determining the MSC yield at first passage. This procedure requires some practice, but has given good yield in large animals (dog, goat, monkey, and pig) as well as man. The bone marrow sample can be processed immediately or stored at room temperature for up to 36 h without detrimental effects. For human marrow samples, a commercial source is Poietics, Inc. (a Cambrex company), which handles all of the informed consent issues, obtains the marrow sample, and delivers it to the investigator for under $1,000.

For small mammals such as rabbits, access to the bone marrow in the long bones is possible by a surgical cutdown, and using relatively stiff tubing placed over the end of the aspiration needle to gain access to marrow further down the bone shaft. For rodents, it is necessary to sacrifice the animal(s), dissect the long bones, remove attached fascia, tendons, and muscles, and move the bones to the sterile tissue culture hood. The bones of up to 10 rats or 20 mice can be processed as a single preparation. The bones are wiped with 70% ethanol to reduce any contamination. The ends of the femur or tibia are removed with sterile wire-cutter-type nippers and the bone marrow is extruded into a collection dish using a needle and saline-filled syringe. The needle should not be too small but should fit easily into the shaft of the femur or tibia bone. The marrow is expressed into a sterile disposable dish, such as a 10-cm tissue culture dish.

2.4 Choices of Growth Media for MSCs

Many different formulations of growth media can be used to isolate and grow MSCs. We have followed the recommendations of Caplan as found in Haynesworth et al. 1992 and use Dulbecco's Modified Eagles media (DMEM) containing 1.0 g/L glucose as opposed to the formulation with 4.5 g/L glucose, which is used in the chondrogenic differentiation media below. Other media that have been used successfully to propagate human MSCs include BGJb, Alpha MEM, DMEM:F12, McCoy's 5A, and RPMI 1640. Although some of these media are richer in certain components than DMEM, they have not proven superior to the original formulation reports and many other works using DMEM.

The other major component of MSC isolation and growth media is fetal bovine serum (FBS). Most media formulations usually use 10% fetal calf serum to provide a mixture of undefined growth factors, cytokines, and attachment factors. It is known that FBS contains platelet-derived growth factor (PDGF), basic fibroblast growth factor (bFGF or FGF-2), and epidermal growth factor (EGF) as well as small amounts of other growth factors. Numerous studies have tested media formulations that included defined growth factors instead of FBS to culture MSCs, but there is not sufficient evidence that these methods produce larger populations of superior MSCs, usually quite the opposite. It is clear that serum-free defined media lack attachment factors to aid MSC attachment and cell yields tend to be low. Some authors have suggested that all media formulations are rendered similar when fetal calf serum is added to the 20% v/v level. In the past, we tested many lots of FBS before selecting one (or two) for purchase. Fetal bovine serum is a by-product of the commodities market, and its price and availability is subject to fluctuations. For that reason it is not always possible to test several lots of serum from different vendors before purchase. For most uses, probably the best solution is to purchase FBS from vendors that market it for use with MSCs or MSC differentiation kits (Cambrex, Inc., Stem Cell Technologies, Inc., Gibco/Invitrogen, Inc.)

2.5 *MSC Growth Medium with 10% FBS and Antibiotics*

1. 445 mL DMEM low glucose (Gibco/Invitrogen) or alternatives Mesencult (Stem Cell Technologies, Inc.) or MSCGM (Cambrex, Inc).
2. 50 mL FCS selected for MSCs (Gibco/ Invtirogen) or alternatives Mesenchymal Stem Cell Stimulatory Supplement (Stem Cell Technologies) or Bullet Kit for MSCs (from Cambrex, Inc).
3. 5 mL Antibiotic/Antimycotic mixture (Gibco/Invitrogen).
4. Mix solutions, filter sterilize and store at 4 °C in the dark for up to 2 wk of use.

2.5.1 Additional Solutions and Items

A number of additional solutions common in tissue culture studies will also be used.

1. Trypsin/EDTA: 0.05% trypsin/0.23 mM EDTA (Gibco/Invitrogen).
2. Dulbecco's phosphate buffered saline (D-PBS) (Gibco/Invitrogen).
3. Ficoll or Percoll Density Gradient solution of 1.073 g/mL (Gibco/Invitrogen).

2.6 Solutions used in Differentiation Assays

1. Adipogenic induction medium (MDI+I medium) consisting of DMEM containing 10% FBS, antibiotics, 0.5 m*M* methyl-isobutylxanthine,1 μ*M* dexamethasone and, 10 μg/mL insulin, 100 μ*M* indomethacin. Use for up to 2 wk.
2. Adipogenic maintenance medium (AM medium) containing 10 μg/mL insulin and 10% FBS in DMEM. Use for up to 2 wk.
3. Chondrogenic differentiation media consists of high glucose (4.5 g/L) DMEM supplemented with 6.25 μg/mL insulin, 6.25 μg/mL transferrin, 6.25 μg/mL selenous acid, 5.33 μg/mL linoleic acid, and 1.25 mg/mL bovine serum albumin (ITS+, Collaborative Research, Cambridge, MA), 0.1 μ*M* dexamethasone, 10 ng/mL TGF-β3, 50 μg/mL ascorbate 2-phosphate, 2 m*M* pyruvate, and antibiotics. All reagents except for TGF-β3 are mixed and the solution can be used for 2 wk, adding TGF-β3 just before applying to cells. TGF-3 should be aliquoted for single use and stored at −80 °C to avoid cycles of freeze/thawing that can cause degradation. The chondrogenic differentiation of rat MSCs is aided by adding 10 ng/mL bone morphogenetic protein 2 or 4 (BMP-2 or BMP-4) to the differentiation medium.
4. Osteogenic supplemented medium (OS medium). DMEM with 10% fetal calf serum with osteogenic supplements 50 μ*M* ascorbate 2-phosphate, 10 m*M* -glycerol phosphate, and 100 n*M* dexamethasone (termed OS medium). This can be made and used for 2 wk.

2.7 Plastic Tissue Culture Wares

MSCs can be grown on a variety of disposable plastic tissue culture containers. We have routinely used products from NUNC and Corning with good success.

3 Methods

3.1 MSC Isolation from Bone Marrow

MSCs can be isolated by a variety of procedures. Below are two useful methods - the direct plating method and the density centrifugation method. Both methods produce very similar populations of MSCs and it is recommended that the novice take the bone marrow tissue sample and divide it, so that one-third is used for the direct plating method and two-thirds is used for the density centrifugation method.

3.2 Direct Plating Method for MSC Isolation

1. Bone marrow sample is diluted with 3 equal volumes of MSC growth medium and distributed equally across several flasks or dishes.
2. Each flask of 75–80 sq cm or 10-cm culture dish receives 10 mL of diluted aspirate.
3. The containers are returned to the 5% CO_2 incubator and cultured undisturbed at 37 °C for 4–5 d.
4. The old medium is aspirated away, without concern for removing the red cells that have settled. Thereafter, the medium is changed every 3–4 d, and contaminating red cells and other nonreplicating and nonattaching cells are eventually diluted and rinsed away.
5. Small MSC colonies of attached fibroblastic cells are visible at 5–7 d. These continue to divide and grow whereas some colonies may not propagate and eventually senesce.
6. After 12–14 d the small colonies are easily found. At this point, the cells are rinsed with serum-free DMEM and subcultured. 5 mL of 0.05 % trypsin/0.23 m*M* EDTA (Gibco/Invitrogen) is added and after several minutes, the cells begin to detach from the substrate. This is observed under the microscope.
7. The excess trypsin/EDTA solution can be carefully aspirated away as long as the cells have not yet fully detached. (If the cells have begun to lift from the substrate, fresh growth medium containing FBS is added and neutralizes the trypsin activity.)
8. The MSCs are rinsed from the surface with growth medium and a pipet and divided in to 2 flasks as at this stage they are at a low concentration but will propagate rapidly. Each flask should contain about 10 mL of cell suspension and are returned to the 5% CO_2 incubator.
9. If needed, the MSCs can be concentrated by centrifugation at 800 g for 5 min at 20 °C in a swinging bucket tabletop centrifuge. The supernatant is aspirated off, leaving less than 0.5 mL over the cell pellet, and new growth medium is added to resuspend the MSCs. The cells are counted using a hemocytometer and placed into flasks at a final concentration of approximately.
10. The human MSCs will continue to grow and approximately every 7 d (4–5 d for other species), they can be subcultured 1:3 with trypsin/EDTA when the cell density is approx 75–80% confluent. Other cell types such as macrophages and fibroblasts either senesce or do not continue to divide in the culture conditions and become diluted out by the propagating MSCs. The MSCs should not be allowed to become confluent as they will become contact inhibited and cease dividing. If they do become confluent, they will begin to divide again when subcultured, but this may change some properties of the cells.

3.3 Density Gradient Isolation Method for MSCs

1. The bone marrow sample is washed by adding 6–8 volumes of Dulbecco's phosphate buffered saline (D-PBS) in a 50 mL disposable centrifuge tube, inverting

gently and subjecting to centrifugation (800 g for 10 min) to pellet the cells to the bottom of the tube.

2. The supernatant is discarded. The cell pellets from all tubes are resuspended in 2–3 mL of D-PBS and combined and an initial cell count performed with a hemocytometer.
3. Up to 4×10^7 cells/mL and 2×10^8 cells total in not more than 10 mL are placed in a 50-mL centrifuge tube and 25 mL of 1.073 g/mL Ficoll or Percoll density solution is added to the bottom of the tube with a pipet. The tubes are subjected to centrifugation at 1,160 *g* for 30 min at room temperature and stopped with the brake off.
4. The upper layer and interface, approx 15–17 mL containing the nucleated cells are collected with a pipet and transferred to a new 50-mL disposable centrifuge tube. Add 2 volumes of D-PBS, cap and mix gently by inversion to wash the cells. The lower layer contains red cells and cell debris and is normally discarded.
5. The tubes with the diluted cells are then subjected to centrifugation at 900 *g* for 5 min at room temperature to pellet the cells with the brake on during deceleration.
6. The supernatant is discarded leaving 0.5–1 mL of D-PBS over the cell pellet. These washed cells are resuspended in ~5 mL of MSC growth medium (DMEM with 10% FBS and antibiotics) and a cell count performed with a hemocytometer. The cells are placed in culture vessels at approx 1.5×10^5 cells per sq cm as shown in Table 2.1 above, and placed in the 5% CO_2 incubator at 37 °C. Colonies of cells will be visible under the microscope in 5–7 d.
7. The cells are ready to subculture in 10–14 d for human cells, 5–7 d for other species, using trypsin/EDTA as described above (begin at step 5). After the first passage the MSCs are normally fed every 3–4 d and subcultured when they reach 75–80 % confluency, approx 1 wk for human MSCs, or 4–5 d for most other species.

3.4 Methods for Freezing and Thawing MSCs

MSCs can be stored for long periods by cryopreservation at −140 °C in growth medium containing 10% DMSO to prevent ice crystal formation. For this,

Table 2.1 Useful seeding densities for MSCs (approximate)

Culture vessel	Surface area in sq cm	Primary MSCs seeding density	Passaged MSCs seeding density	Medium volume
12-well	3.8	6.2×10^5	2.1×10^4	1.5 mL
6-well	9.6	1.6×10^6	5.0×10^4	2.0 mL
100 cm	55.0	9.0×10^6	3.0×10^5	10.0 mL
80 T Flask	80.0	1.3×10^7	4.3×10^5	15.0 mL
185 T Flask	185.0	3.0×10^7	1.0×10^6	35.0 mL

1. Remove log phase MSCs from the substrate by trypsin/EDTA as for propagation, centrifuge the cells for 5 min at 800 *g* to concentrate, aspirate the supernatant leaving the cell pellet and less than 0.5 mL of medium.
2. Add freezing medium and gently mix to achieve a cell concentration of 2×10^6 to 10×10^6 MSCs/mL. The vials are frozen to −80 °C overnight and then transferred to −140 °C (liquid nitrogen vapor phase).
3. To thaw for use, a vial of MSCs is removed from the −140 °C freezer and rapidly thawed in a 30–37 °C water bath. The vial is moved through the water to encourage thawing, and the vial is removed from the water bath when just a small amount of ice remains.
4. The vial is wiped with 70% ethanol and open in the biological safety cabinet. The contents are transferred to a flask or culture dish containing at least 10 volumes of growth medium and the culture vessel is placed in the 5% CO_2 incubator. The medium is replaced with fresh growth medium after 24–48 h.

   1) Alternatively, the thawed cells can be transferred to a large disposable centrifuge tube containing 10 volumes of growth medium and the cells are subjected to centrifugation at 800 *g* for 5 min. The supernatant is aspirated away and the cell pellet is resuspended in fresh growth medium and transferred to the culture vessel. The growth medium is replaced at 3 d.

3.5 Flow Cytometry Evaluation of MSCs

Cell populations grown in culture are often characterized to evaluate the surface molecules that represent the receptors for extracellular matrix components, growth factors and cell-cell interactions. Fluorescence activated cell sampling (FACS) analysis is a valuable and convenient method for determining surface molecules on MSCs. A number of the known surface molecules found on MSCs are listed in Table 2.2, although none of these by themselves can be said to identify MSCs or any other stem cell. In general, although surface molecules can be damaged by proteases such as trypsin, the surface molecules on MSCs have not shown a difference when trypsin is used to release them from the substrate, in a manner similar to that performed during the subculturing process.

1. Harvest MSCs with 0.05% trypsin, 25 m*M* EDTA in DPBS. Centrifuge at 800 *g* for 5 min at room temperature to collect the MSCs.
2. Count the cells with the hemocytometer and dilute the MSCs to 0.5×10^6 cells/mL.
3. Approx 100,000–200,000 MSCs in DPBS are stained for 20 min at room temperature with 10–20 µL of antibody as determined from the manufacturer's recommendation and a preliminary test run.
4. The MSCs are then washed 2 times with 5 volumes of DPBS, with centrifugation at 800 *g* for 5 min at room temperature.
5. The labeled MSCs are finally resuspended in Flow Buffer containing 1% formaldehyde, 0.1% sodium azide and 0.5 %bovine serum albumin in DPBS.

Table 2.2 Surface antigen detection on MSCs[a]

CD11a,b	–
CD13	+
CD14	–
CD18 Integrin β2	–
CD29	+
CD31 PECAM	–
CD34	–
CD44	+
CD45	–*
CD49b Integrin α2	+
CD49d Integrin α4	–
CD49e Integrin α5	+
CD50 ICAM3	–
CD54 ICAM1	+
CD62E E-Selectin	–
CD71 Transferrin Rec	+
CD73 SH-3	+
CD90 Thy-1	+
CD105 Endoglin,	+
CD106 VCAM	+
CD117	–
CD133	–
CD166 ALCAM	+
Nestin	+
p75 LNGR	+
HLA ABC	+
HLA DR	inducible with IFN
SSEA 3, 4	+
TRK (ABC)	+
Differentiation to	
Adipo	+
Osteo	+
Chondro	+
Stromal	+
Neural	(+)
Myoblast Sk	(+)
Endothelial	(+)

[a]Note: + Positive, – Negative, (+) Detection varied, * Positive on isolation, lost in culture

6. The cells are analyzed on a flow cytometer such as a Guava PC (Gauva, Inc.) or FACS Caliber (Becton-Dickinson, Inc), collecting 10,000 events and analysis by Cell Quest (BD) or other software package.

After isolation from bone marrow and by the second passage in culture, the human MSCs will show very good reproducibility in their flow cytometry results from passage to passage, and donor to donor.

4 Methods of Inducing MSC differentiation

Methods to demonstrate differentiation of MSCs to 3 different lineages were first presented in 1996 at the American Society for Cell Biology annual meeting and published in greater detail in 1999 (*4*, see web figures). As a fourth lineage, the MSCs produce many growth factors and cytokines and can also serve as stromal cells to support growth of HSCs *(4,15)* and even human embryonic stem cells *(33)*. Additional molecular techniques can be used to demonstrate appropriate gene expression for each of the different lineages as presented previously *(4,11–22)*. A summary of the MSC differentiation is presented in Fig. 2.1.

4.1 Adipogenic Differentiation of MSCs

The method described here can be found in Pittenger et al. (*4*, see web figures) and in greater detail in Patent #6,322,784 available at www.uspto.gov/patft/index.html.

1. To induce adipogenic differentiation, MSCs are cultured as monolayers in DMEM, 10% FBS with antibiotics and allowed to become confluent as adipogenic differentiation does not occur in sub-confluent cultures.
2. The cells are cultured for 3–7 d more and the growth medium is aspirated off and replaced with adipogenic induction medium (MDI+I medium) consisting of DMEM containing 10% FBS, antibiotics, 1 μ*M* dexamethasone and 0.5 m*M* methyl-isobutylxanthine, 10 μg/mL insulin, 100 μ*M* indomethacin.
3. The hMSCs are incubated in this media for 48–72 h and the media then changed to adipogenic maintenance medium (AM medium) containing 10 μg/mL insulin and 10% FBS in DMEM for 24 h. The cells were then retreated with MDI+I for a second 3 d treatment followed by 24 h in AM media, and finally a third treatment with MDI+I.
4. The cultures are then maintained in AM for 1 wk to allow the lipid vacuoles to accumulate within the cytoplasm. Therefore this process takes up to 2 weeks, but signs of adipogenic differentiation are evident after the first incubation in MDI+I.
5. The lipid vacuoles are seen in more and more cells over the course of MDI+I treatment, and the vacuoles will be seen to coalesce within the adipocyte over time.
6. Nile Red staining is used to quantify lipid vacuoles as described in Pittenger et al.(4, see web figures). Nile Red is a vital dye and easily taken up by cells in 15 min and accumulates in the lipid vacuoles, making them very visible under the fluorescence microscope in the green channel. DAPI can be use to counterstain the nuclei.
7. Nile Red (Sigma #N3013) is dissolved in acetone at 1 mg/mL and added to the culture medium at a final concentration of 5 μg/mL. The dye accumulates in lipid vesicles in about 15 min and does not seem to harm the living cells.

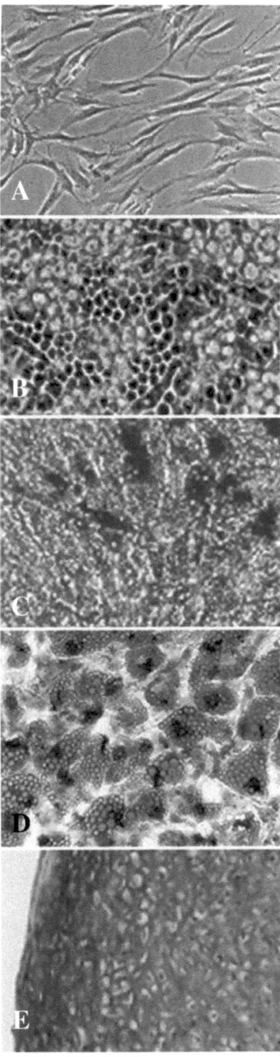

Fig. 2.1 Differentiation of Human MSCs. MSCs isolated from the bone marrow taken from the iliac crest as described were subjected to differentiation conditions as described. Top to bottom (**A**) MSCs in log phase culture have a spindle- shaped or fibroblastic appearance. (**B**) Hematopoietic cells growing on top of a feeder layer of human MSCs have a "cobblestone" appearance. As the MSCs are contact inhibited as they reach confluence, there is no need to irradiate the feeder cells. (**C**) MSCs under osteogenic conditions for 7 d that have been stained for alkaline phosphatase activity and also stained by the von Kossa method to label the calcium deposits with silver. (**D**) Adopogenic MSCs after the recommended 3 treatments with MDI+I and several days in adipose maintenance medium to allow the lipid vacuoles to grow large. The culture has been stained with oil red O and hematoxylin as a counter stain. (**E**) Chondrogenic differentiation of MSCs is clearly revealed by collagen type II staining in the extracellular matrix. The MSCs were cultured as a micromass pellet in serum-free medium containing TGFβ3 for 3 wk and then prepared for histology in paraffin sections. The sections have been stained with a monoclonal antibody to collagen II *(4,40)* and counterstained with hematoxylin

8. The results can easily be read on a fluorescence plate reader to quantify approximate cell number by DAPI in the blue channel and lipid accumulation by Nile Red staining in the green channel (4, see web figures). Alternatively, the cells can be fixed with 3.7% formaldehyde and the lipid deposits within the cells can be stained with oil red O, and counterstained with fast green or hematoxylin. Methanol based solvents however may leach out the oil red O.

4.2 Chondrogenic Differentiation of MSCs

The development of an in vitro chondrogenic assay took a long period, perhaps because it was necessary to remove the fetal bovine serum that seemed so important for cell isolation and proliferation, and to wait a relatively long period for differentiation to occur. The method described here can be found in greater detail in Mackay et al *(34)*.

1. Chondrogenic differentiation media consists of high glucose (4.5 g/l) DMEM supplemented with 6.25 μg/mL insulin, 6.25 μg/mL transferrin, 6.25 μg/mL selenous acid, 5.33 μg/mL linoleic acid, 1.25 mg/mL bovine serum albumin (ITS+, Collaborative Research, Cambridge, MA), 0.1 μ*M* dexamethasone, 10 ng/mL TGF-β3, 50 μg/mL ascorbate 2-phosphate, 2 m*M* pyruvate, and antibiotics. TGF-β3 should be aliquoted for single use and stored at −80 °C to avoid cycles of freeze/thawing that can cause degradation. The chondrogenic differentiation of rat MSCs is aided by adding 10 ng/mL bone morphogenetic protein 2 or 4 (BMP-2 or BMP-4) to the differentiation medium *(21)*.
2. Chondrogenic differentiation is induced by placing 2.5×10^5 MSCs into the defined chondrogenic medium and subjecting the cells to gentle centrifugation (800 *g* for 5 min) in a 15 mL conical polypropylene tube, after which, the cap is loosened and the tube placed in the incubator where the cells adhere to one another and consolidate into a cell pellet within 24 h.
3. After several days, it will be noticed that the MSCs form a 1 mm ball in the bottom of the tube. The chondrogenic medium made with fresh TGFβ every 3-4 days and the medium changed by *careful* aspiration, as the cell pellets are free floating. If the cell pellets are found attached to the tube wall, they are gently dislodged with a pipet tip.
4. Chondrogenic differentiation of MSCs can take several weeks, although in weeks 2–3 the pellets often begin to show signs of enlargement owing to extracellular matrix accumulation.
5. Routinely, several samples are prepared in parallel such that a cell pellet can be harvested at week 1, 2, 3, and 4.
6. At harvest, a variety of methods can be used to analyze differentiation. For histology, the pellets were fixed in 4% formaldehyde, paraffin embedded, sectioned and analyzed by immunostaining for collagen II expression, etc. Sections can also be stained with Safranin O for detection of proteoglycans.

7. If RNA is to be isolated for gene expression studies, it will likely take 4–5 cell pellets to isolate enough RNA for evaluation.
8. Generally the week-1 and week-2 samples show limited differentiation, whereas the 3- and 4-week samples offer extensive chondrogenic differentiation. The chondrogenic differentiation of rat MSCs is aided by adding 10 ng/mL bone morphogenetic protein 2 or 4 (BMP-2 or BMP-4) to the differentiation medium *(21)*.

4.3 Osteogenic Differentiation of MSCs

1. MSCs are seeded into 6 well culture plates at a low density of 3×10^4 cells per well in MSC growth medium.
2. After 24 h, this medium was replaced with the same medium with osteogenic supplements 50 μM ascorbate 2-phosphate, 10 mM -glycerol phosphate, and 100 nM dexamethasone (termed OS medium).
3. As the MSCs proliferate in the OS medium, they can be seen to aggregate and this "nodule formation" is evident under the microscope.
4. The nodules are areas of high alkaline phosphatase activity, and the wells can be stained to reveal the increased activity. For this, cultures are usually stained at 7 d after addition of OS medium using a commercially available kit such as Sigma #85L-2.
5. Mineralization of the nodules is usually evidenced by dense, refractile deposits. The amount can be assayed and quantified by measuring calcium deposition in the well using chemical methods such as the Stanbio Laboratory Calcium Liquicolor kit. (available from Cambrex, Inc).
6. Mineralization can also be demonstrated by the increase in calcium deposits as revealed by silver deposition using the Von Kossa staining method (*4*, see web figures, 13, 15, 19, and 36).

5 Conclusions

The MSC isolation and characterization methods described in this chapter have proven useful over a number of years, and in a variety of species, and provide a starting point for a variety of other studies. The MSCs found in bone marrow can be isolated by simple standardized methods and culture expanded to billions of cells. The propagated MSCs retain many useful properties for cellular therapy including the production of growth factors and cytokines, the ability to differentiate to specific lineages and cell types, the ability to be transplanted and engrafted in new sites, the ability to avoid immune rejection and the ability to modulate immune responses. MSCs have already provided a wealth of new information on our understanding of adult progenitor cells that will prove useful for the development of cellular therapies for a variety of tissues damaged by trauma, disease, or genetic insufficiency. Many of the methods and findings developed with MSCs will prove useful for cellular therapies involving other cell types including, eventually, human embryonic stem cells, or cell types developed by therapeutic cloning.

References

1. Caplan, A. I (1991) Mesenchymal stem cells. *J. Orthop. Res.* 9:641–650.
2. Haynesworth, S. E., Baber M. A., and Caplan, A. I. (1992) Cell surface antigens on human marrow-derived mesenchymal cells are detected by monoclonal antibodies. *Bone* 13:69–80.
3. Lazarus, H. M., Haynesworth, S. E., Gerson, S. L., Rosenthal, N. S., and Caplan, A. I. 1995. Ex vivo expansion and subsequent infusion of human bone marrow-derived stromal progenitor cells (mesenchymal progenitor cells): implications for therapeutic use. *Bone Marrow Transplant*16(4):557–564.
4. Pittenger, M. F., Mackay, A. M., Beck, S. C., Jaiswal, R. K., Douglas, R., Mosca, J. D., Moorman, M. A., Simonetti, D. W., Craig, S., and Marshak, D. R. (1999) Multilineage potential of adult human mesenchymal stem cells. *Science* 284:143–147.
5. Friedenstein, A. J., Petrakova, K. V., Kurolesova, A. I., and Frolova, G. P. (1968) Heterotopic transplants of bone marrow: Analysis of precursor cells for osteogenic and haematopoietic tissues. *Transplantation* 6:230–247.
6. Friedenstein, A. J. (1976) Precursor cells of mechanocytes. *Int. Rev. Cytol.* 47:327–355.
7. Owen, M. E. and Friedenstein, A. J. (1988) Stromal stem cells: marrow derived osteogenic precursors. *Ciba Found. Symp.* 136:42–60.
8. Halvorsen, Y. C., Wilkison, W. O., and Gimble, J. M. (2000) Adipose-derived stromal cells-their utility and potential in bone formation. *Int J Obes Relat Metab Disord.* 24:S41–44.
9. Gronthos, S., Franklin, D. M., Leddy, H. A., Robey, P. G., Storms, R. W., and Gimble, J. M. (2001) Surface protein characterization of human adipose tissue-derived stromal cells. *J Cell Physiol.* 189:54–63.
10. Zuk, P. A., Zhu, M., Ashjian, P., De Ugarte, D. A., Huang, J. I., Mizuno, H., Alfonso, Z. C., Fraser, J. K., Benhaim, P., and Hedrick, M. H. (2002) Human adipose tissue is a source of multipotent stem cells. *Mol Biol Cell.*13:4279–4295.
11. Lodie, T. A., Blickarz, C. E., Devarakonda, T. J., He, C., Dash, A. B., Clarke, J., Gleneck, K., Shihabuddin, L., and Tubo, R. (2002) Systematic analysis of reportedly distinct populations of multipotent bone marrow-derived stem cells reveals a lack of distinction. *Tissue Eng.* 8:739–751.
12. Phinney, D. G., Kopen, G., Isaacson, R. L., and Prockop, D. (1999) Plastic adherent stromal cells from the bone marrow of commonly used strains of inbred mice: variations in yield, growth and differentiation. *J. Cell. Biochem.* 72:570–585.
13. Saito, T., Dennis, J. E., Lennon, D. P., Young, R. G., and Caplan, A. I. (1995) Myogenic expresssion of mesenchymal stem cells within myotubes of mdx mice in vitro and in vivo. *Tissue Engin.* 1(4): 327–343.
14. Pereira, R. F., Halford, K. W., O'Hara, M. D., Leeper, D. B., Sokolov, B. P., Pollard, M. D., Bagasra, O., and Prockop, D. J. (1995) Cultured adherent cells from marrow can serve as long-lasting precursor cells for bone, cartilage and lung in irradiated mice. *Proc. Natl. Acad. Sci. USA.* 92:4857–4861.
15. Dennis, J. E., Merriam, A., Awadalla, A., Yoo, J. U., Johnstone, B., and Caplan, A. I. (1999) A quadripotent mesenchymal progenitor cell isolated from the marrow of an adult mouse. *J. Bone Miner. Res.* 14:700–709.
16. Short, B., Brouard, N., Driessen, R., and Simmons, P. J. (2001). Prospective isolation of stromal progenitor cells from mouse BM. *Cytotherapy.* 3:407, 408.

17. Grigoriadis, A. E., Heersche, J., and Aubin, J. E. (1990) Continuously growing bipotential and monopotential myogenic, adipogenic and chondrogenic subclones isolated from the multipotential RCJ3.1 clonal cell line. *Developmental Biology* 142:313–318.
18. LeBoy, P. S., Beresford, J., Devlin, C., and Owen, M. (1991) Dexamethasone induction of osteoblast mRNAs in rat marrow stromal cell cultures. *J. Cell Physiol.* 146:370–378.
19. Beresford, J. N., Bennett, J. H., Devlin, C., LeBoy, P. S., and Owen, M. (1992) Evidence for an inverse relationship between the differentiation of adipocytic and osteogenic cells in rat marrow stromal cell cultures. *J. Cell Sci.* 102:341–351.
20. Kadiyala, S., Jaiswal, N., and Bruder, S. P. (1997) Culture-expanded bone marrow-derived mesenchynal stem cells can regenerate a critical-sized segmental bone defect. *Tissue Engin.* 3:173–185.
21. Neuhuber, B., Gallo, G., Howard, L., Kostura, L., Mackay, A., and Fischer, I. (2004) Reevaluation of in vitro differentiation protocols for bone marrow stromal cells: disruption of actin cytoskeleton induces rapid morphological changes and mimics neuronal phenotype. *J. Neurosci. Res.* 77(2):192–204.
22. Wakitani, S., Goto, T., Pineda, S. J., Young, R. G., Mansour, J. M., Goldberg, V. M., and Caplan, A. I. (1994) Mesenchymal cell based repair of large, full thickness defects of articular cartilage. *J. Bone Joint Surg.* 76:579–592.
23. Grande, D. A., Southerland, S. S., Ryhanna, Manji, B. S., Pate, D. W., Schwartz, R. E., and Lucas, P. A. (1995) Repair of articular defects using mesenchymal stem cells. *Tissue Engin.* 1:345–353.
24. Young, R. G., Butler, D. L., Weber, W., Caplan, A. I., Gordon, S. L., and Fink, D. J. (1998) Use of mesenchymal stem cells in a collagen matrix for achilles tendon repair. *J. Orthop. Res.* 16:406–413.
25. Kadiyala, S., Young, R. G., Thiede, M. A., and Bruder, S. P. (1997) Culture expanded canine mesenchymal stem cells possess osteochondrogenic potential in vivo and in vitro. *Cell Transplant.* 6(2):125–134.
26. Kraitchman, D. L., Tatsumi, M., Gilson, W. D., Ishimori, T., Kedziorek, D., Walczak, P., Segars, W. P., Chen, H. H., Fritzges, D., Izbudak, I., Young, R. G., Marcelino, M., Pittenger, M. F., Solaiyappan, M., Boston, R. C., Tsui, B. M., Wahl, R. L., and Bulte, J. W. (2005) Dynamic imaging of allogeneic mesenchymal stem cells trafficking to myocardial infarction. *Circulation.* 112(10):1451–1461.
27. Murphy, J. M., Fink, D. J., Hunziker, E. B., and Barry, F. P. (2003). Stem cell therapy in a caprine model of osteoarthritis. *Arthritis Rheum.* 48:3464–3474.
28. Shake, J. G., Gruber, P. J., Baumgartner, W. A., Senechal, G., Meyers, J., Redmond, J. M., Pittenger, M. F., and Martin, B. J. (2002) Mesenchymal stem cell implantation in a swine myocardial infarct model: engraftment and functional effects. *Ann. Thorac. Surg.* 73:1919–1925.
29. Kraitchman, D. L., Heldman, A. W., Atalar, E., Amado, L. C., Martin, B. J., Pittenger, M. F., Hare, J. M., and Bulte, J. W. (2003). In vivo magnetic resonance imaging of mesenchymal stem cells in myocardial infarction. *Circulation.* 107(18):2290–2293.
30. Amado, L. C., Saliaris, A. P., Schuleri, K. H., St John, M., Xie, J. S., Cattaneo, S., Durand, D. J., Fitton, T., Kuang, J. Q., Stewart, G., Lehrke, S., Baumgartner, W. W., Martin, B. J., Heldman, A. W., and Hare, J. M. (2005) Cardiac repair with intramyocardial injection of allogeneic mesenchymal stem cells after myocardial infarction. Proc. Natl Acad. Sci. USA. 102(32):11,474–11,479.
31. Freyman, T., Polin, G., Osman, H., Crary, J., Lu, M., Cheng, L., Palasis, M., and Wilensky, R. L. (2006) A quantitative, randomized study evaluating three methods of mesenchymal stem cell delivery following myocardial infarction. *Eur. Heart J.* 27(9):1114–1122.
32. Mahmud, N., Pang, W., Cobbs, C., Alur, P., Borneman, J., Dodds, R., Archambault, M., Devine, S., Turian, J., Bartholomew, A., Vanguri, P., Mackay, A., Young, R., and Hoffman, R. (2004) Studies of the route of administration and role of conditioning with radiation on unrelated allogeneic mismatched mesenchymal stem cell engraftment in a nonhuman primate model. *Exp. Hematol.* 32(5):494–501.

33. Chapel, A., Bertho, J. M., Bensidhoum, M., Fouillard, L., Young, R. G., Frick, J., Demarquay, C., Cuvelier, F., Mathieu, E., Trompier, F., Dudoignon, N., Germain, C., Mazurier, C., Aigueperse, J.. Borneman, J., Gorin, N. C., Gourmelon, P., and Thierry, D. (2003). Mesenchymal stem cells home to injured tissues when co-infused with hematopoietic cells to treat a radiation-induced multi-organ failure syndrome. *J. Gene Med.* 5(12):1028–1038.
34. Devine, S. M., Cobbs, C., Jennings, M., Bartholomew, A., and Hoffman, R. (2003) Mesenchymal stem cells distribute to a wide range of tissues following systemic infusion into nonhuman primates. *Blood.* 101(8):2999–3001.
35. Bartholomew, A., Patil, S., Mackay, A., Nelson, M., Buyaner, D., Hardy, W., Mosca, J., Sturgeon, C., Siatskas, M., Mahmud, N., Ferrer, K., Deans, R., Moseley, A., Hoffman, R., and Devine, S. M. (2001) Baboon mesenchymal stem cells can be genetically modified to secrete human erythropoietin in vivo. *Hum. Gene Ther.* 12(12):1527–1541.
36. Haynesworth SE, Baber MA, and Caplan AI (1996) Cytokine expression by human marrow-derived mesenchymal progenitor cells in vitro: effects of dex and IL-1α. *J. Cell. Physiol.* 166: 585-592.
37. Majumdar, M. K., Thiede, M. A., Mosca, J. D., Moorman, M., and Gerson, S. L. (1998). Phenotypic and functional comparison of marrow-derived mesenchymal stem cells and stromal cells. *J. Cell Phys.* 176:57–66.
38. Reese, J. S., Koc, O. N., and Gerson, S. L. (1999) Human mesenchymal stem cells provide stromal support for efficient CD34+ transduction. *J. Hematother. Stem Cell Res.* 8:515–523.
39. Cheng, L., Hammond, H., Ye, Z., Zhan, X., and Dravid, G. (2003) Human adult marrow cells support prolonged expansion of human embryonic stem cells in culture. *Stem Cells.* 21(2):131–142.
40. Mackay, A. M., Beck, S. C., Murphy, J. M., Barry, F. P., Chichester, C. O., and Pittenger, M. F. (1998) Chondrogenic differentiation of cultured human mesenchymal stem cells from marrow. *Tissue Eng.* 4(4):415–428.
41. Johnstone, B., Hering, T. M., Caplan, A. I., Goldberg, V. M., and Yoo, J. U. (1998) In vitro chondrogenesis of bone marrow-derived mesenchymal progenitor cells. *Exp. Cell Res.* 10;238(1):265–272.
42. Jaiswal, R. K., Jaiswal, N., Bruder, S. P., Mbalaviele, G., Marshak, D. R., and Pittenger, M. F. (2000). Adult human mesenchymal stem cell differentiation to the osteogenic or adipogenic lineage is regulated by mitogen-activated protein kinase. *J. Biol. Chem.* 275(13):9645–9652.
43. Bittira, B., Shum-Tim, D., Al-Khaldi, A., and Chiu, R. C. 2003. Mobilization and homing of bone marrow stromal cells in myocardial infarction. *Eur. J. Cardiothorac. Surg.* 24(3): 393–398.
44. Pittenger, M. F., and Martin, B. J. (2004) Mesenchymal stem cells and their potential as cardiac therapeutics. *Circ. Res.* 95(1):9–20.
45. Francois, S., Bensidhoum, M., Mouiseddine, M., Mazurier, C., Allenet, B., Semont, A., Frick, J., Sache, A., Bouchet, S., Thierry, D., Gourmelon, P., Gorin, N. C,, and Chapel, A. (2006). Local irradiation not only induces homing of human mesenchymal stem cells at exposed sites but promotes their widespread engraftment to multiple organs: a study of their quantitative distribution after irradiation damage. *Stem Cells.* 24:1020–1029.
46. Klyushnenkova, E., Mosca, J. D., Zernetkina, V., Majumdar, M. K., Beggs, K. J., Simonetti, D. W., Deans, R. J., and McIntosh, K. R. (2005) T cell responses to allogeneic human mesenchymal stem cells: immunogenicity, tolerance, and suppression. *J. Biomed. Sci.* 12(1):47–57.
47. Di Nicola, M., Carlo-Stella, C., Magni, M., Milanesi, M., Longoni, P. D., Matteucci, P., Grisanti, S., and Gianni, A.M. (2002) Human bone marrow stromal cells suppress T-lymphocyte proliferation induced by cellular or nonspecific mitogenic stimuli, *Blood.* 99 (2002), 3838–3843.
48. Aggarwal, S., and Pittenger, M. F. (2005) Human mesenchymal stem cells modulate allogeneic immune cell responses. *Blood.* 105(4):1815–1822.
49. Le Blanc, K., and Pittenger, M. F. (2005) Mesenchymal stem cells: progress toward promise. *Cytotherapy.* 7(1):36–45.

50. Groh, M. E., Maitra, B., Szekely, E., and Koc, O.N. (2005) Human mesenchymal stem cells require monocyte-mediated activation to suppress alloreactive T cells. *Exp. Hemat.* 33(8):928–934.
51. Le Blanc, K., Rasmusson, I., Sundberg, B., Gotherstrom, C., Hassan, M., Uzunel, M., and Ringden, O. (2004) Treatment of severe acute graft-versus-host disease with third party haploidentical mesenchymal stem cells. *Lancet.* 363(9419):1439–1441.
52. Horwitz, E. M., Gordon, P. L., Koo, W. K., Marx, J. C., Neel, M. D., McNall, R. Y., Muul, L., and Hofmann, T. (2002) Isolated allogeneic bone marrow-derived mesenchymal cells engraft and stimulate growth in children with osteogenesis imperfecta: Implications for cell therapy of bone. *Proc. Natl Acad. Sci. USA*.99(13):8932–8937.
53. Le Blanc, K., Gotherstrom, C., Ringden, O., Hassan, M., McMahon, R., Horwitz, E., Anneren, G., Axelsson, O., Nunn, J., Ewald, U., Norden-Lindeberg, S., Jansson, M., Dalton, A., Astrom, E., and Westgren, M. (2005) Fetal mesenchymal stem-cell engraftment in bone after in utero transplantation in a patient with severe osteogenesis imperfecta. *Transplantation.* 79(11):1607–1614.
54. Koc, O. N., Day, J., Nieder, M., Gerson, S. L., Lazarus, H. M., and Krivit, W. (2002) Allogeneic mesenchymal stem cell infusion for treatment of metachromatic leukodystrophy (MLD) and Hurler syndrome (MPS-IH). *Bone Marrow Transplant.* 30(4):215–222.

Chapter 3
A Method to Isolate and Purify Human Bone Marrow Stromal Stem Cells

Stan Gronthos and Andrew C. W. Zannettino

Abstract The STRO-1 antibody can be used as a single reagent to isolate human bone marrow stromal stem cells (BMSSC), owing to its restricted specificity to a cell surface molecule expressed by clonogenic BMSSC, with little or no reactivity to hematopoietic stem/progenitor populations or mature stromal elements. The present protocol uses a combination of two different immunoselection methodologies in an attempt to generate highly purified preparations of BMSSC. This process involves the initial isolation of a minor subpopulation of bone marrow mononuclear cells (approx 10%) expressing the STRO-1 antigen, by means of magnetic activated cell sorting. Dual-color fluorescence activated cell sorting is then used as a secondary step to further purify the rare STRO-1^{bright} expressing fraction that contains all of the colony-forming BMSSC, based on their co-expression of a secondary cell surface marker, CD106 (VCAM-1).

Keywords Bone marrow stromal stem cells; mesenchymal stem cells; magnetic activated cell sorting; fluorescence activated cell sorting.

1 Introduction

The bone marrow (BM) connective tissue and surrounding bone are believed to arise from a pool of self-replicating multipotential stromal stem cells (BMSSC) that may be similar to other mesenchymal stemlike cells described in different tissues *(1–3)*. These plastic adherent adult stem/progenitor cells derived from bone marrow were originally referred to as fibroblastoid colony forming units (CFU-F), then in the hematological literature as marrow stromal cells, subsequently as mesenchymal stem cells, and most recently as multipotent mesenchymal stromal cells (MSCs). However, the identification and purification of BMSSC has been largely restricted, in part, owing to their low incidence in aspirates of human marrow (0.05–0.001%) *(1,4–6)*, and to a lack of specific antibody reagents that allow for the precise identification and selective isolation of BMSSC. While, early studies relied on the capacity of human BMSSC to adhere to plastic in the presence of high serum levels *(1,7)*, this rather crude method of isolating and propagating BMSSC has made it difficult to unravel the important factors and

From: *Methods in Molecular Biology, vol. 449, Mesenchymal Stem Cells: Methods and Protocols*
Edited by D.J. Prockop, D.G. Phinney, and B.A. Bunnell © Humana Press, Totowa, NJ

conditions that are critical for colony formation, cellular growth, and development in the absence of accessory cells and different serum components.

The recent development of novel antibodies recognizing antigens present on the cell surface of BMSSC, that are correspondingly not reactive with hematopoietic progenitors, has led to an increased understanding of the biological properties of BMSSC. Thus far, various antibody reagents, reactive with different antigens including STRO-1, CD18, CD49a/CD29, NGF-R, PDGF-R, EGR-R, IGF-R, CD106, CD146, and HOP-26, have been used, with varying efficiencies, to positively select for BMSSC from aspirates of human bone marrow *(4,5,8–17)*. Comparative tests have demonstrated that the STRO-1 antibody has the highest affinity and efficiency for isolating all clonogenic BMSSC as a stand alone reagent (unpublished observations). Further development of stromal specific monoclonal reagents that identify discrete developmental stages may provide essential reagents to enable further characterization of the cellular properties and functions of the BMSSC population. The present chapter describes a method for generating highly purified preparations of BMSSC from human BM, based principally on their high expression of the STRO-1 antigen *(4,16)*.

2 Materials

2.1 Processing of Bone Marrow Mononuclear Cells (BMMNC)

1. Sodium heparin (Fisons Pharmaceuticals, Australia).
2. Ficoll-Hypaque, Lymphoprep, 1.077 g/dL (Nycomed Pharma AS, Oslo, Norway).
3. Hanks balanced salt solution (HBSS; JRH Biosciences, Lenexa, KS, USA).
4. HEPES, pH 7.35 (JRH Biosciences).
5. PBS, phosphate buffered saline solution, pH 7.4 (JRH Biosciences).
6. Blocking buffer; HHF supplemented with 5% (v/v) normal human serum.
7. Fetal bovine serum (FBS; JRH Biosciences, Lenexa, KS, USA).
8. HHF; HBSS supplemented with 20 m*M* HEPES, pH 7.35 and 5% (v/v) FBS.
9. 70-μm Falcon cell strainer (Becton Dickinson Biosciences).
10. 50-mL Falcon tube (Becton Dickinson Biosciences, San Jose, CA, USA).
11. 14-mL polystyrene Falcon tubes (Becton Dickinson Biosciences).
12. White Cell Fluid; 2% acetic acid in distilled H_2O.

2.2 Magnetic Activated Cell Sorting and Fluorescence Activated Cell Sorting

1. Biotinylated goat anti-mouse IgM, μ-chain specific, (Southern Biotechnology Associates, Birmingham, UK).

2. Goat anti-mouse IgG, γ-chain specific, conjugated to phycoerythrin (PE), (Southern Biotechnology Associates, Birmingham, UK).
3. Streptavidin conjugated fluorescein isothiocyanate (FITC), (Caltag Laboratories, San Francisco, CA, USA).
4. Streptavidin microbeads, (Miltenyi Biotec; Bergisch Gladbach, Germany).
5. Streptavidin LS Columns, (Miltenyi Biotec).
6. 4- and 14-mL polypropylene Falcon tubes (Becton Dickinson Biosciences).
7. FACS Fix; 1% (v/v) formalin, 0.1 *M* D-glucose, 0.02% sodium azide in PBS).
8. MACS buffer (Ca^{2+} - and Mn^{2+} -free PBS was supplemented with 1% BSA in PBS, 5 m*M* EDTA and 0.01% sodium azide).
9. Antibodies used: Mouse isotype-matched negative controls (murine IgM and IgG_1,; Caltag, Burlingame, CA); STRO-1 (murine IgM anti-human stromal cell; Developmental Studies Hybridoma Bank, University of Iowa, Iowa City, IA, USA); 6G10 (murine IgG_1 anti-CD106, VCAM-1; American Type Culture Collection Manassas,VA. ATCC No. HB 10519). Working dilutions: monoclonal supernatants (1/2) and purified antibodies (10 μg/mL).

2.3 Cell Culture of Human BMSSC

1. α-Modification of Eagle's Medium (α-MEM; JRH Biosciences).
2. Bovine serum albumin (Cohn fraction V; Sigma-Aldrich Corp., St Louis, MO).
3. Recombinant human platelet derived growth factor-BB (PDGF), (CytoLab Ltd., Rehovot, Israel).
4. L-Ascorbate-2-phosphate (WAKO Pure Chemical Industries, Japan).
5. L-glutamine (200 m*M* stock; JRH).
6. Penicillin (5,000 U/mL) and streptomycin (5,000 μg/mL) 100× stock solution (JRH Biosciences).
7. Recombinant human insulin (Sigma-Aldrich Corp).
8. Human low density lipoprotein (Sigma-Aldrich Corp).
9. Iron saturated human transferrin (Sigma-Aldrich Corp).
10. Dexamethasone sodium phosphate (DEX: David Bull Laboratory, Sydney, Australia).
11. β-mercaptoethanol (BDH Chemicals, Poole, UK).
12. 0.5% trypsin/EDTA solution (JRH Biosciences).
13. T-25 and T-75 culture flasks (CellStar, Greiner Bio-One, Frickenhausen, GmbH).
14. 6-well culture plates (NUNC, Intermed, Roskilde, Denmark).
15. 0.22-μm Ministart low protein binding filters (Sartorius, Goettingen, Germany).
16. 1.8-mL Cryo-tubes (Nalge Nunc International, Rochester, NY, USA).
17. Dimethyl sulphoxide (DMSO; BDH Chemicals).

3 Methods

The methods described below outline (1) the isolation and preparation of the bone marrow mononuclear cells (BMMNC), (2) the magnetic activated cell separation (MACS) of clonogenic BMSSC or CFU-F, (3) methods to enumerate the efficiency of the MACS enrichment, (4) methods to ex vivo culture, expand and cryopreserve human BMSSC grown under serum or serum-deprived conditions.

3.1 Preparation of Human Bone Marrow (BM) Mononuclear Cells

3.1.1 Collection of Human Bone Marrow (BM) and Preparation of BM Mononuclear Cells by Density gradient separation

1. Following informed consent, collect approx 40 mL of human bone marrow (BM) from health young volunteers (18–40 y) by aspiration from the posterior iliac crest (hip bone). BM should be placed immediately into a preservative-free, sodium heparin-containing 50-mL tube (10,000 units/tube), (*see* **Note 1**).
2. Remove a 10-μL aliquot and dilute 1:20 (i.e., add to 190 μL) into White Cell Fluid (WCF) and enumerate nucleated cell content using a hemocytometer (*see* **Notes 2 and 3**).
3. An equal volume of blocking buffer is then added to the BM aspirate, mixed well, the strained through a 70-μm Falcon cell strainer to remove any small clots and bone fragments.
4. Dispense 3 mL of Ficoll-Hypaque (Lymphoprep) solution in the bottom of approx 12 round bottom 14-mL polystyrene Falcon tubes and carefully overlay with 7.5 mL of diluted BM.
5. Centrifuge tubes at 400 *g* for 30 min at room temperature. *Keep centrifuge brake off.*
6. Using a disposable plastic Pasteur pipet, recover the leucocyte band from all tubes and pool into 4 × 14 mL polypropylene tubes.
7. Dilute cells with HHF wash buffer and pellet the BMMNC by centrifugation of the sample at 400 *g* for 10 min at 4 °C. *Keep centrifuge break on.*
8. Aspirate the buffer and repeat step 7 until all cells are pooled into one tube.

3.2 Magnetic Activated Cell Sorting (MACS) of STRO-1 Positive BMSSC

The use of MACS allows for partial purification of the BMSSC population the processing of large numbers of BMMNC without compromising high losses in overall stem cells yield. Following density gradient centrifugation, we routinely recover approx 1–2 × 10^8 mononuclear cells from a BM aspirate of 40 mL. Before immunolabeling, BMMNC are resuspended in 0.5 mL blocking buffer and incubated on ice for approx 30 min to reduce the possibility of Fc receptor-mediated binding of antibodies.

3.2.1 Assessment of Bone Marrow Quality by Colony-Efficiency Assay

Details of this procedure have been described previously *(12,18)*. The expected incidence of CFU-F colony in human bone marrow aspirates is approx 5–10 CFU-F per 10^5 cells plated. Lower CFU-F frequencies can indicate high levels of peripheral blood contamination.

1. The BMMNC are seeded into 6-well culture plates at 0.3, 1.0, and 3.0 × 10^5 cells per well in α-MEM supplemented with 20% (v/v) FBS, 2 m*M* L-glutamine, 100 μ*M* L-ascorbate-2-phosphate, 50 U/mL penicillin, 50 mg/mL streptomycin, and β-mercaptoethanol (5 × 10^{-5} *M*). Cultures are set up in triplicate and incubated at 37 °C in 5% CO_2 and >90% humidity for 12 d (*see* **Note 4**).
2. Day 12 cultures are washed twice with PBS and then fixed for 20 min in 1% (w/v) paraformaldehyde in PBS.
3. The fixed cultures can then be stained with 0.1% (w/v) toluidine blue (in 1% paraformaldehyde solution) for 1 h then rinsed with tap water and allowed to dry. Aggregates of greater than 50 cells are scored as CFU-F using an Olympus SZ-PT dissecting light microscope (Olympus Optical Co. Ltd., Tokyo, Japan). Colonies should be visually checked at day 10 to ensure that there is no overgrowth of cells making it difficult to enumerate individual colonies.

3.2.2 Isolation of STRO-1$^+$ BMSSC Using Magnetic Activated Cell Sorting (MACS)

1. Pellet BMMNC by centrifugation at 400 *g* at 4 °C for 10 min and resuspend in 500 μl of STRO-1 supernatant per 5 × 10^7 BMMNC and incubate on ice for 60 min with occasional, gentle mixing (*see* **Note 5**).
2. BMMNC are washed twice in HHF wash buffer and then resuspended in 0.5 mL of HHF containing biotinylated goat anti-mouse IgM (μ-chain specific) at a 1/50 dilution and incubated at 4 °C for 45 min.
3. The BMMNC are washed three times in MACS buffer (*see* **Note 6**) and resuspended in 450 μL of MACS buffer to which 50 μL of streptavidin microbeads added (10 μL of microbeads/10^7 cells in 90 μL MACS buffer). The mixture is incubated on ice for 15 min.
4. After one wash in ice-cold MACS buffer, a small aliquot of cells is removed for flow cytometric analysis (pre sample). The remaining cells are then placed onto the mini MACS column (column capacity of 10^8 cells, Miltenyi Biotec, MS column).

The STRO-1^{-} cells (negative fraction) are not retained within the column and pass through into a fresh 2 mL polypropylene tube, under gravity into the effluent, while the STRO-1^{+} cells remain attached to the magnetised matrix.

5. Wash the column 3 times with 0.5 mL MACs buffer to remove any nonspecifically bound STRO-1^{-} cells, which are collected in a fresh 2 mL polypropylene tube. Apply gentle positive pressure with plunger if the flow rate stops prematurely.
6. The STRO-1^{+} cells (positive fraction) are recovered by flushing the column with MACS buffer into a fresh 2-mL polypropylene tube after withdrawing the column from the magnetic field. The STRO-1^{+} cells are then counted and processed for two-color FACS as described below in Subsection 3.2.
7. Small samples (0.5–1.0 × 10^5 cells) from each of the pre-MACS, STRO-1^{-}, and STRO-1^{+} fractions are removed into separate 2-mL polypropylene tubes containing 0.2 mL of streptavidin-FITC conjugate (1/50). The cell samples are then incubated for an additional 5 min on ice to enable assessment of the enrichment procedure. A sample of (1.0 × 10^5 cells) unlabeled pre-MACS cells can be used as a negative control.
8. These samples are washed twice in HHF, fixed in FACS Fix solution and subsequently analysed by flow cytometry to assess purity and recovery. An example of which is shown in Fig. 3.1.
9. At this point, the partially purified STRO-1^{+} BMSSC can be culture expanded as described in subsection 3.4 or further purified by two-color FACS as described below (Subsection 3.2).
10. The remaining STRO-1^{-} population can be used to further isolate other cell populations such as CD34^{+} hematopoietic stem cells.

3.3 Fluorescence Activated Cell Sorting of Highly Purified BMSSC

While, all measurable CFU-F are contained within the STRO-1^{+}BMMNC fraction, BMSSC only represent less than 2% of the total STRO-1^{+} population. The majority of the STRO-1^{+} cells are glycophorin-A^{+} nucleated red cells and some CD19^{+} B-cells *(5)*. Therefore, the selection of BMSSC based on STRO-1 expression alone results in only a partial enrichment of CFU-F (approx 10-fold) *(12)*. Recent work in our laboratory has shown that clonogenic BMSSC are all contained within the STRO-1^{bright} cell fraction that can be further discriminated by dual-color FACS based on the expression of markers that are absent on nucleated red cells and lymphocytes, for example CD106 and CD146 *(4,16)*. The methods described below enable the isolation of a minor subpopulation of the total STRO-1^{+} cell fraction, STRO-1^{bright}/CD106^{+} BMMSC (1.4% ± 0.3; *n* = 20), in which 1 in every 2–3 cells plated have the capacity to form a CFU-F *(4,16)*. This level of enrichment is almost 5,000-fold higher than the average incidence of CFU-F observed with unfractionated BMMNC (1 CFU-F per 10,000 cells plated), (*see* **Note 7**).

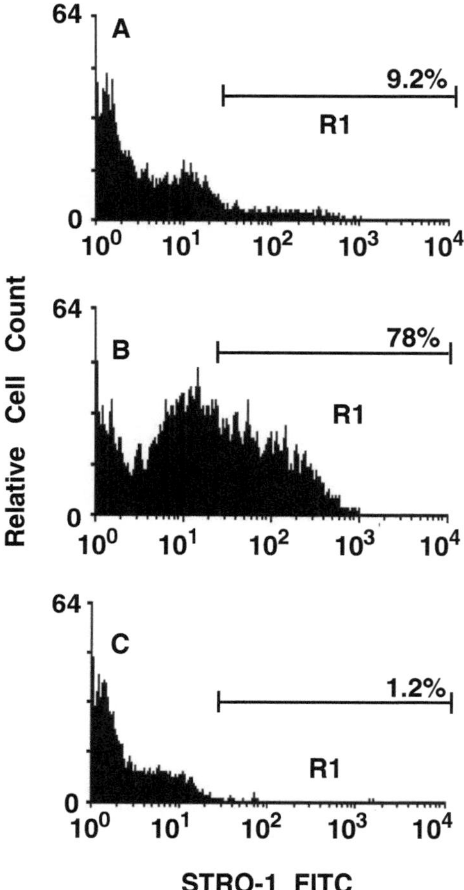

Fig. 3.1 Purity check of MACS-isolated STRO-1^+ BMMNC by flow cytometric analysis. (**A**) The representative histogram shows the level of STRO-1 expression in the BMMNC population before MACS-isolation (pre-MACS) relative to the level of fluorescence of less than 1% in a sample of unlabeled BMMNC, region 1 (R1). (**B**) The level of purity obtained for the STRO-1^+ BMMNC population (**B**) retained on the magnetized column is greatly enhanced following one round of MACS-isolation. (**C**) The relative low level STRO-1 expression in BMMNC after passing through the magnetized column (STRO-1^-) is also shown. If low levels of enrichment for the STRO-1^+ BMMNC population are obtained (less than 50%) then the STRO-1^+ fraction can be passed through a second magnetized column to obtain a higher level of purity

3.3.1 Isolation of STRO-1^{bright}/CD106^+ BMSSC Using Flow Cytometric Cell Sorting (FACS)

1. Before immunolabeling, the MACS-isolated STRO-1^+ cell BMMNC (routinely $2–5 \times 10^6$ cells-from 1×10^8 BMMNC) are resuspended in 0.5 mL HHF in preparation for 2-color immunofluorescence (refer to Fig. 3.2) and FACS.

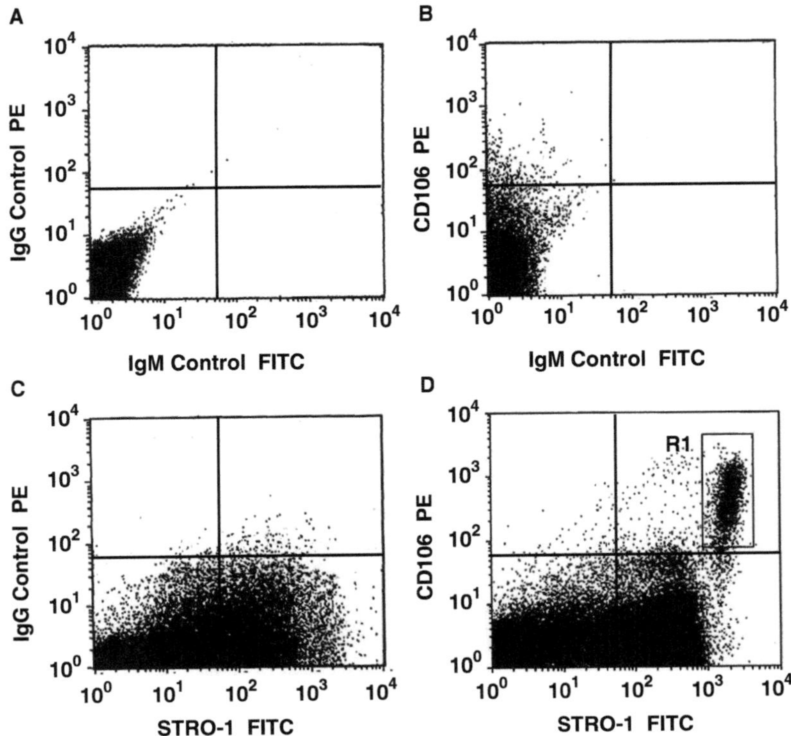

Fig. 3.2 Purification of human BMSSC by two-color FACS. Dot plots A to C are used for compensation purposes only. (**A**) Flow cytometric analysis of unlabeled MACS-isolated STRO-1 BMMNC (double negative control). (**B**) Flow cytometric analysis of MACS-isolated STRO-1 BMMNC labeled with IgG anti-human CD106 (VCAM-1) then PE-conjugated goat anti-mouse IgG, (PE control). (**C**) Flow cytometric analysis of MACS-isolated STRO-1 BMMNC labeled with streptavidin conjugated-FITC (FITC control). (**D**) Flow cytometric analysis of MACS-isolated STRO-1 BMMNC labeled with IgG anti-human CD106 (VCAM-1) then PE-conjugated goat anti-mouse IgG and streptavidin conjugated-FITC, (sample tube). The clonogenic BMSSC population resides in the STRO-1^{bright}/CD106$^+$ cell fraction as indicated in region 1 (R1)

2. Approx 3–5 × 10^5 MACS-isolated STRO-1^+ cell are dispensed into 3 appropriately labeled tubes, to which the following are added:
3. (i) No primary antibody (double negative control), kept on ice.
   (ii) Streptavidin-FITC conjugate (1/100 dilution in HFF) incubated on ice for 30 min (FITC control). The cells are then washed twice in HHF.
   (iii) 0.5 mL of murine IgG anti-human CD106 (VCAM-1) diluted to 20 µg/mL in HFF. The STRO-1^+ cells are incubated on ice for 30 min, washed twice in HHF and resuspended in 0.2 mL of PE-conjugated goat anti-mouse IgG (γ-chain specific), (PE control). The sample is incubated and washed as before then resuspended in HFF.

(iv) The remaining 1–2 × 10^6 MACS-isolated STRO-1^+ cells are resuspended in 0.5 mL murine IgG anti-human CD106 (VCAM-1) and incubated as above, washed twice in HHF and resuspended in 0.2 mL of PE-conjugated goat anti-mouse IgG (γ-chain specific) and Streptavidin-FITC conjugate (1/100 dilution in HHF), then incubated on ice for 30 min (sorting sample). The cells are then washed as before then resuspended in HHF.

4. The samples are resuspended at a concentration of 1 × 10^7 cells per mL in HHF before sorting on any sorter fitted with a 250 MW argon laser emitting light at a wavelength of 488 nm able to simultaneously detect FITC and PE. Samples (i–iii) are used to establish compensation for both FITC and PE as shown in Fig. 3.3. Ensure the high expressing STRO-1^+ cells are observable by reducing the voltage for the FITC channel.
5. Sorted STRO-1^{bright}/VCAM-1^+ cells from sample (iv) are collected in tubes containing appropriate growth media and mixed. *Do not resuspend the samples in growth media, which may cause froth to form in the flow cytometric sorter.*
6. A cell count is performed as described above. The sorted cells are then cultured as described in Subsection 3.4.

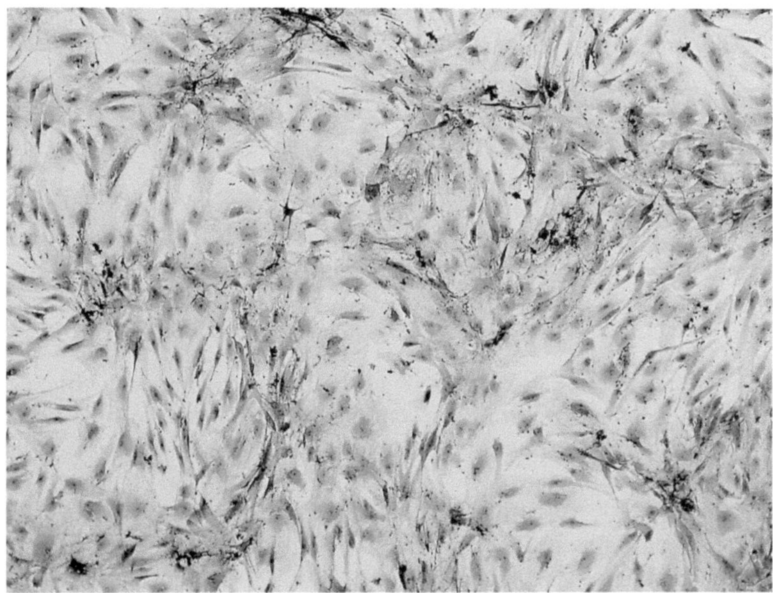

Fig. 3.3 Ex vivo expanded human BMSSC. The picture depicts a representative field of ex vivo expanded BMSSC at 90% confluency, and is the recommended cell density before subculture (100×). The adherent cells were fixed with 0.1% (w/v) toluidine blue in 1% paraformaldehyde solution. Hyperconfluent, over grown BMSSC cultures can lead to cell aggregation and clumping following trypsin/EDTA digestion

3.4 Ex Vivo Culture of Human BMSSC

3.4.1 Serum Replete Medium

1. The STRO-1^{bright}/CD106^{+} isolated BMSSC populations (at 1–3 × 10^4 per cm^2) are cultured in tissue culture flasks or plates containing α-modification of Eagle's Medium (α-MEM) supplemented with 20% foetal bovine serum, 100 μ*M* L-ascorbate-2-phosphate, 2 m*M* L-glutamine, 50 U/mL penicillin and 50 μg/mL streptomycin at 37 °C in 4% CO_2 at relative humidity of >90% for 2 wk. Primary BMSSC populations should be passaged when the cultures achieve 80–90% confluency (Fig. 3.2).
2. Adherent cultures are washed 1× with serum free HBSS and the cells liberated by enzymatic digestion by the addition of 2 mL of 0.5% Trypsin/EDTA solution per T75 flask for 5–10 min at 37°C (*see* **Note 8**).
3. Cell viability is assessed by preparing a 1:5 dilution of single cell suspension in 0.4% trypan blue/PBS, and the number of viable cells (dead cells take up the blue dye) determined using a haemocytometer.
4. BMMSC single cell suspensions are pooled and re-seeded at 0.5–1.0 × 10^4 per cm^2 in α-MEM growth medium supplemented with 10% FBS, 100 μ*M* L-ascorbate-2-phosphate, 2 m*M* L-glutamine, 50 U/mL penicillin and 50 μg/mL streptomycin and incubated at 37 °C in 5% CO_2 at relative humidity of >90%. Cultures are fed twice weekly by aspirating out the medium and replacing with an equal volume of freshly prepared medium warmed to 37 °C (*see* **Note 9**).

3.4.2 Serum Deprived Medium

This method is a modification of the serum deprived medium (SDM) developed initially for the growth of hematopoietic progenitor cells *(19,20)* and was adapted for the growth of BMSSC *(12,18)*. Similar serum deprived growth conditions have also been shown to support the growth of multipotential adult progenitor cells (MAPC) *(21,22)*.

1. Prepare fibronectin coated plates or flasks by precoating with 5 μg per cm^2 fibronectin solution for 90 min at room temperature. After this, the fibronectin solution is aspirated off and the culture vessels washed once with sterile PBS before seeding with cells.
2. The STRO-1^{bright}/CD106^{+} isolated BMSSC populations (at 1–3 × 10^4 per cm^2) are cultured in the fibronectin-coated tissue culture flasks or plates suspended in media containing α-MEM supplemented with 2% (w/v) bovine serum albumin (Cohn fraction V), 10 μg/mL recombinant human insulin, human low density lipoprotein, 200 μg/mL iron saturated human transferrin, 2 m*M* L-glutamine, dexamethasone sodium phosphate (10^{-8} *M*), 100 μ*M* L-ascorbic acid-2-phosphate, β-mercaptoethanol (5 × 10^{-5} *M*), 10 ng/mL platelet derived growth factor-BB, 50 U/mL penicillin and 50 μg/mL streptomycin (*see* **Note 10**).

3. The cultures are then incubated at 37 °C in 4% CO_2 at relative humidity of >90% for 2 wk. Primary BMSSC populations should be passaged when the cultures achieve 80–90% confluency as described above (*see* **Notes 11 and 12**).

3.4.3 Cryopreservation of Ex Vivo Expanded MPC

1. Routinely, single cell suspensions of culture expanded MPC are prepared by trypsin/EDTA digest as described above. The cells are then diluted and washed in cold HFF.
2. The cell pellet is resuspended at a concentration of 1×10^7 cells per mL in FBS and maintained on ice. An equal volume of freeze mix (20% DMSO in cold FBS) is then added gradually while gently mixing the cells to give a final concentration of 5×10^6 cells per mL in a 10% DMSO/FBS. One-milliliter aliquots are then distributed into 1.8-mL cryovials precooled on ice, i.e., 1 mL per tube, then frozen at a rate of –1 °C per minute using a rate control freezer (*see* **Note 13**).
3. The frozen vials are then transferred to liquid nitrogen for long-term storage. Recovery of the frozen stocks is achieved by rapid thawing the cells in a 37 °C water bath. The cells are then resuspended in cold HFF and spun at 280*g* for 10 min.
4. To assess viability of the cells, prepare a 1:5 dilution in 0.4% trypan blue/PBS, and the number of cells determined using a haemocytometer. Typically this procedure gives viabilities between 80–90%.

4 Notes

1. BMMNC from healthy volunteers can also be purchased commercially (e.g., Poietic Technologies, Gaithersburg, MD, USA).
2. Wipe excess marrow off the pipet tip with a tissue to ensure that that the cell number is not over estimated.
3. Alternate methods of cell enumeration may be used (e.g., using an automated Coulter™ Counter).
4. Variations in FBS batches can severely hamper establishment of CFU-F colonies and growth. Batch testing of FBS is highly recommended to ensure optimal growth conditions.
5. The STRO-1 antibody can also be purchased commercially from R&D Systems Inc., Minneapolis, MN, USA.
6. The MACS buffer should be de-gassed before use by loosening the bottle's lid then placing the bottle half-filled with buffer into a sealed chamber under vacuum for 1 to 2 h.
7. In addition to CD106, costaining of STRO-1^{bright} cells can also be achieved using antibodies reactive to other markers (e.g., CD18, CD90, CD146, NGF-R, PDGF-R, EGR-R, and IGF-R).
8. Adherent cell cultures if over confluent can be liberated using collagenase/dispase digestion. Wash the cells with PBS then incubate in a prewarmed solution of collagenase (3 mg/mL in PBS) (collagenase type I; Worthington Biochemical Co., Freehold, NJ) and dispase (4 mg/mL in PBS) (Neutral Protease Grade II; Boehringer Mannheim, GMBH, Germany) (1 mL per 25 cm^2 surface area) for 60 min at 37 °C. Single cell suspensions were then washed twice in HHF buffer.
9. Growth medium can be premade, sterilized using a 0.22-μm filter, and then stored at 4 °C. If the medium is greater than a week old at 4 °C then fresh 2 m*M* L-glutamine should be added before use. Prior to culture, if cells appear to be clumping pass them through a 70-μm cell strainer.

10. The growth of CFU-F under serum-deprived conditions can also be achieved in the presence of EGF (10 ng/mL) alone or a combination of both PDGF-BB and EGF (10 ng/mL).
11. Serum deprived medium should be made fresh at time of use and sterilized using a 0.22-μm filter. The different media components should be stored in small aliquots at −80 °C.
12. Following establishment of primary BMSSC cultures in serum deprived conditions, precoating of flasks with fibronectin is no longer necessary for initiating secondary BMSSC cultures owing to endogenous extracellular matrix production.
13. The BMSSC can also be frozen using a Cryo 1 °C freezing container "Mr. Frosty" (Nalge Nunc International, Rochester, NY, USA) by placing the container holding the cryotubes at −80 °C overnight before transferring the cells into liquid nitrogen. For serum free applications ProFreeze solution (CAMBREX Bio Science, Walkersville, MD, USA) containing a final concentration of 7.5% DMSO can be substituted for 10%FBS/DMSO freeze mix.

Acknowledgments The authors wish to acknowledge the technical assistance of Ms. Fiona Kohr and Ms. Sharon Paton. This work was supported by National Health and Medical Council of Australia grants (A Zannettino, S Gronthos).

References

1. Castro-Malaspina, H., Gay, R. E., Resnick, G., Kapoor, N., Meyers, P., Chiarieri, D., McKenzie, S., Broxmeyer, H. E., and Moore, M. A. (1980) Characterization of human bone marrow fibroblast colony-forming cells (CFU-F) and their progeny. Blood 56(2):289–301.
2. Friedenstein, A. J., Chailakhjan, R. K., and Lalykina, K. S. (1970) The development of fibroblast colonies in monolayer cultures of guinea- pig bone marrow and spleen cells. Cell Tissue Kinet 3(4):393–403.
3. Owen, M., and Friedenstein, A. J. (1988) Stromal stem cells: marrow-derived osteogenic precursors. Ciba Found Symp 136(29):42–60.
4. Gronthos, S., Zannettino, A. C., Hay, S. J., Shi, S., Graves, S. E., Kortesidis, A., and Simmons, P. J. (2003) Molecular and cellular characterisation of highly purified stromal stem cells derived from human bone marrow. J Cell Sci 116(Pt 9):1827–1835.
5. Simmons, P. J., and Torok-Storb, B. (1991) Identification of stromal cell precursors in human bone marrow by a novel monoclonal antibody, STRO-1. Blood 78(1):55–62.
6. Waller, E. K., Olweus, J., Lund-Johansen, F., Huang, S., Nguyen, M., Guo, G. R., and Terstappen, L. (1995) The "common stem cell" hypothesis reevaluated: human fetal bone marrow contains separate populations of hematopoietic and stromal progenitors. Blood 85(9):2422–2435.
7. Castro-Malaspina, H., Rabellino, E. M., Yen, A., Nachman, R. L., and Moore, M. A. (1981) Human megakaryocyte stimulation of proliferation of bone marrow fibroblasts. Blood 57(4):781–787.
8. Boiret, N., Rapatel, C., Veyrat-Masson, R., Guillouard, L., Guerin, J. J., Pigeon, P., Descamps, S., Boisgard, S., and Berger, M. G. (2005) Characterization of nonexpanded mesenchymal progenitor cells from normal adult human bone marrow. Exp Hematol 33(2):219–225.
9. Deschaseaux, F., Gindraux, F., Saadi, R., Obert, L., Chalmers, D., and Herve, P. (2003) Direct selection of human bone marrow mesenchymal stem cells using an anti-CD49a antibody reveals their CD45med,low phenotype. *Br J Haematol* **122**(3):506–517.
10. Filshie, R. J., Zannettino, A. C., Makrynikola, V., Gronthos, S., Henniker, A. J., Bendall, L. J., Gottlieb, D. J., Simmons, P. J., and Bradstock, K. F. (1998) MUC18, a member of the immunoglobulin superfamily, is expressed on bone marrow fibroblasts and a subset of hematological malignancies. *Leukemia* **12**(3):414–421.

11. Gronthos, S., Graves, S. E., Ohta, S., and Simmons, P. J. (1994) The STRO-1+ fraction of adult human bone marrow contains the osteogenic precursors. *Blood* **84**(12):4164–4173.
12. Gronthos, S., and Simmons, P. J. (1995) The growth factor requirements of STRO-1-positive human bone marrow stromal precursors under serum-deprived conditions in vitro. *Blood* **85**(4):929–940.
13. Joyner, C. J., Bennett, A., and Triffitt, J. T. (1997) Identification and enrichment of human osteoprogenitor cells by using differentiation stage-specific monoclonal antibodies. *Bone* **21**(1):1–6.
14. Miura, M., Chen, X. D., Allen, M. R., Bi, Y., Gronthos, S., Seo, B. M., Lakhani, S., Flavell, R. A., Feng, X. H., Robey, P. G., Young, M., and Shi, S. (2004) A crucial role of caspase-3 in osteogenic differentiation of bone marrow stromal stem cells. *J Clin Invest* **114**(12):1704–1713.
15. Quirici, N., Soligo, D., Bossolasco, P., Servida, F., Lumini, C., and Deliliers, G. L. (2002) Isolation of bone marrow mesenchymal stem cells by anti-nerve growth factor receptor antibodies. *Exp Hematol* **30**(7):783–791.
16. Shi, S., and Gronthos, S. (2003) Perivascular niche of postnatal mesenchymal stem cells in human bone marrow and dental pulp. *J Bone Miner Res* **18**(4):696–704.
17. Simmons, P. J., Gronthos, S., Zannettino, A., Ohta, S., and Graves, S. (1994) Isolation, characterization and functional activity of human marrow stromal progenitors in hemopoiesis. *Prog Clin Biol Res* **389:**271–280.
18. Gronthos, S., Graves, S. E., and Simmons, P. J. (1998) Isolation, purification and in vitro manipulation of human bone marrow stromal precursor cells. In: Beresford JNaO, M.E. (ed.) Marrow stromal cell culture. Cambridge University Press, Cambridge, pp. 26–42.
19. Lansdorp, P. M., and Dragowska, W. (1992) Long-term erythropoiesis from constant numbers of CD34+ cells in serum-free cultures initiated with highly purified progenitor cells from human bone marrow. *J Exp Med* **175**(6):1501–1509.
20. Migliaccio, G., Migliaccio, A. R., and Adamson, J. W. (1988) In vitro differentiation of human granulocyte/macrophage and erythroid progenitors: comparative analysis of the influence of recombinant human erythropoietin, G-CSF, GM-CSF, and IL-3 in serum-supplemented and serum-deprived cultures. *Blood* **72**(1):248–256.
21. Reyes, M., Dudek, A., Jahagirdar, B., Koodie, L., Marker, P. H., and Verfaillie, C. M. (2002) Origin of endothelial progenitors in human postnatal bone marrow. *J Clin Invest* **109**(3):337–346.
22. Reyes, M., and Verfaillie, C. M. (2001) Characterization of multipotent adult progenitor cells, a subpopulation of mesenchymal stem cells. *Ann N Y Acad Sci* **938:**231–233; discussion 233–235.

Chapter 4
Adipose-Derived Stem Cells

John K. Fraser, Min Zhu, Isabella Wulur, and Zeni Alfonso

Abstract Human adipose tissue has been shown to contain a population of cells that possesses extensive proliferative capacity and the ability to differentiate into multiple cell lineages. These cells are referred to as adipose tissue-derived stem cells (ADSCs) and are generally similar, though not identical, to mesenchymal stem cells (also referred to as marrow stromal cells). ADSCs for research are most conveniently extracted from tissue removed during an elective cosmetic liposuction procedure but may also be obtained from resected adipose tissue. This chapter describes surgical procedures associated with improved ADSC recovery and the processes by which aspirated adipose tissue is washed and digested with collagenase to yield a heterogeneous population from which ADSCs can be expanded. The large volume of tissue obtained from a liposuction procedure (average ~2 L), combined with the relatively high frequency of ADSC within the digestate, yields substantially more stem cells than can be realized from marrow without extensive expansion in culture.

Keywords Adipose tissue; mesenchymal stem cell; adipose tissue-derived stem cell; liposuction; CFU-F.

1 Introduction

The term mesenchymal stem cell refers to the plastic adherent adult stem/progenitor cells from bone marrow originally referred to as fibroblastoid colony forming units, then in the hematological literature as marrow stromal, subsequently as mesenchymal stem cells, and most recently as multipotent mesenchymal stromal cells (MSCs). However, although MSC are the best-described population of cells that exhibit extensive proliferative capacity and the ability to generate progeny of the connective tissue lineages (bone, cartilage, tendon, fat, etc), other cells have been shown to exhibit a similar phenotype. In 2001 Zuk et al. described multilineage differentiation from a population of cells derived by enzymatic digestion of human adipose tissue *(1)*. This work was followed by studies using clonally derived populations

From: *Methods in Molecular Biology, vol. 449, Mesenchymal Stem Cells: Methods and Protocols*
Edited by D.J. Prockop, D.G. Phinney, and B.A. Bunnell © Humana Press, Totowa, NJ

demonstrating that multilineage differentiation was the property of single cells within the population *(2)*; an observation that has now been confirmed in both human *(3)* and murine *(4)* adipose tissue-derived cells. These adipose tissue-derived stem cells (ADSC; also referred to as adipose-derived adult stem cells *(5)*, adipose-derived stromal cells *(6)*, and human multipotent adipose-derived stem cells *(7,8)*) are characterized by extensive proliferative capacity and multilineage differentiation. In our hands and those of others, ADSC express CD105/endoglin, CD44, CD90/Thy1, and SH2 but are negative for expression of CD45 and CD31 *(2,6,9)*. However, differences in the expression of certain surface molecules including VCAM-1 and VLA-4 have been reported *(9)* as have differences in the ability to differentiate towards the osteogenic and chondrogenic lineages *(10,11)* and in basal gene expression *(12)*. Further, we have reported that ADSC are less stringent than MSC in their requirement for prescreened lots of sera for their growth *(1,2)*. This may reflect the relative frequency of the stem cells within the starting population; MSC, as measured by clonogenic cell assays are generally reported at a frequency of between 1 in 50,000 and 1 in 1 million in the marrow of skeletally mature adults *(13–17)* whereas digestion of adult human adipose tissue releases buoyant adipocytes yielding an ADSC frequency in the nonbuoyant fraction of between 1 in 1,000 and 1 in 30 (see the following) *(18,19)*.

The relative frequency of clonogenic cells is one reason for interest in adipose tissue as a stem cell source. Another is the relative ease and low morbidity with which adipose tissue can be harvested. The American Society for Aesthetic Plastic Surgery reported that 478,251 people underwent cosmetic liposuction in the USA during 2004 *(20)*. This represents an enormous volume of immediately available tissue that can be obtained following a very simple informed consent process, without risk to the donor provided that due attention is paid to protecting privacy. The purpose of the present chapter is to describe the methods by which this material can be processed to yield ADSC for research purposes.

2 Materials

1. Sterile saline for tissue washing.
2. Dulbecco's phosphate buffered saline supplemented with 25 m*M* HEPES, and 2% human serum albumin or fetal calf serum.
3. Type IA Collagenase (Sigma, Catalog #C-2674). Other commercially available collagenase preparations such as Roche's Blendzyme family also provide satisfactory results.
4. Sterile separatory funnels with stop cock.
5. Shaking water bath or incubator.
6. DMEM/F-12 50/50,1× with L-glutamine (Mediatech Catalog # 10-090-CV).
7. Antibiotic/Antimycotic 100X solution (ABAM), sterile, contains 10,000 units Penicillin-G/mL;10,000 mcg Streptomycin/mL; 25 mcg Amphotericin B/mL (Omega Scientific Catalog # AA-40).

8. Complete culture medium: DMEM/F-12 based commercially prepared cell culture media in which 10% fetal bovine sera (FBS) and 1% v/v antibiotic/antimycotic have been previously added.
9. Falcon 3046 Multiwell 6-well tissue culture plate, Becton Dickinson.
10. 10% formalin.
11. Hematoxylin (Gill III formula, Surgipath Medical Industries, Cat# 01542).

3 Methods

3.1 Tissue Harvest

Liposuction is generally performed in a process in which tumescent solution, a mixture of saline, epinephrine (added as a vasoconstrictor to reduce blood loss), and lidocaine (for local anesthesia), is injected into the subcutaneous space *(21,22)*. Most commonly a 2–5-mm diameter steel cannula is then inserted through a small (0.5 cm) incision in the skin and rapidly and repeatedly moved within the space to disrupt the tissue (a process referred to as *tunneling*). Tissue is then aspirated through holes in the cannula into a collection trap. Variations on this general theme include; use of manual aspiration via syringe for small volume tissue removal (usually <100 mL), use of powered cannula in which a reciprocating motion increases tissue disruption, and the application of ultrasound energy through the cannula, which further disrupts tissue and adipocytes. Further, in many procedures the surgeon will change the cannula multiple times to obtain greater control over the new body contour. Clearly, for ethical reasons the researcher can have no impact on the approach chosen by surgeon and patient, yet these variables can have substantial impact on the quality of the ADSC-containing population. Generally speaking, the less energy applied to the tissue, the greater the viability of the final product. Thus, manual or simple suction-assisted lipoplasty using a large diameter (>3 mm) cannula will yield tissue that has been subject to less shear force within the system than high-powered, machine-assisted suction through a small diameter cannula. By the same token, we have made anecdotal observations that ultrasound-assisted lipoplasty is associated with a substantial (>70%) decrease in the yield of ADSC. Thus, in preparing for research using human liposuction-derived material it is essential to evaluate the approach most frequently applied by the surgeon potentially providing access to tissue and to select those surgeons whose practice population and clinical preferences provide the most suitable aspirate. Because of subtleties in individual practice (for example, dose and use of epinephrine, cannula diameter, applied suction force, etc.), we suggest that investigators initially work with a number of different surgeons to build a sense of which surgeons provide the most reliable material.

Tissue may also be obtained following informed consent during an unrelated procedure such as hip surgery or resection of excess skin, or by small volume, syringe-mediated suction of research volunteers under local anesthesia.

3.2 *Tissue Processing*

Aspirated tissue is generally collected into a sealed nonsterile container that is disposed of as biohazardous medical waste. The material collected is a mixture of tumescent solution, blood, free lipid released from lysed adipocytes, and aspirated tissue fragments the precise proportions of which are largely determined by physician practice. Following informed consent this material should be transported to the laboratory for processing as quickly as possible; we have noted that the yield of ADSC, as measured by the fibroblast colony-forming unit (CFU-F) assay, falls by approx 50% for 24 h of storage before initiation of processing. Because the common collection containers are not designed for transport, precautions must be taken to avoid contamination and spills during transfer from the surgical facility to the processing laboratory. We suggest bagging in a sealed, spill-proof pouch, placing this bag within a rupture-proof secondary container such as a Tyvek® bag, and finally use of a crush-resistant outer container bearing labeling consistent with local and federal regulations.

Human lipoaspirate is frequently obtained without the researchers being aware of the infectious disease status of the donor; that is, without available results of serologic testing for biohazardous agents such as human immunodeficiency virus (HIV) and hepatitis B. For this reason all procedures involving manipulation of tissue should be performed in a protective environment such as a biological safety cabinet and operators should wear appropriate protective clothing and equipment at all times during tissue processing.

3.2.1 Tissue Washing

The buoyancy of adipose tissue is such that processing can be performed using approaches similar to those used in organic chemistry. Specifically, we apply sterile (autoclaved) separator funnels in tissue processing to separate buoyant tissue fragments from tumescent solution and blood. Washing may be repeated until the buoyant fraction is a vivid orange color and the infranatant is clear. Use of large volume funnels allows maximization of the ratio of saline:aspirate and more efficient washing. Alternatively, washing may be performed in beakers and the infranatant removed by aspiration.

1. Place stopcock to the closed position and decant lipoaspirate into the sterile separatory funnel.
2. Add sterile saline, prewarmed to 37 °C, and invert the funnel 4–5 times with the cap in place. Return to the upright position and allow 3–5 min for phase separation.
3. Remove the cap, open the stopcock and let blood-saline mixture flow into a liquid pathological waste container. Close the stopcock before the fat-blood/saline interface.
4. Repeat steps 2 and 3 until the infranatant is clear or residual opacity no longer declines substantially with additional wash cycles.

3.2.2 Tissue Digestion

A number of different enzymes and enzyme combinations have been described for digestion of human adipose tissue *(2,23,24)*. We have developed a proprietary mixture of enzymes (Celase®) that optimizes processing. However, off-the-shelf enzymes such as the collagenase preparations listed above will yield satisfactory results.

1. Estimate the volume of washed fat (volume of fat after the last wash).
2. Prepare an equal volume of warm, sterile buffered saline containing 500 CDU/mL (equivalent to 0.5 Wünsch units/mL) collagenase.
3. Pour washed fat from the separatory funnel into a 600 mL, 1,000 mL, or 2,000 mL sterile bottle, depending on the estimated volume of washed fat (container volume should be at least 4 times that of the aspirate).
4. Add the buffered saline/collagenase mixture, seal the container and place on a thermal shaker, prewarmed to 35–38 °C for 20 ± 5 min. Initiate shaking.
   a. The frequency and amplitude of shaking should be set such that it is just sufficient to prevent separation of the buoyant tissue from the collagenase solution. Excessive amplitude or frequency can cause loss of cell recovery.
5. Inspect the digestion frequently after the first 15 min to ensure that over-digestion does not occur. Digestion time will vary with different tissue donors and physicians. For example, a larger cannula may generate larger fragments of tissue that may take longer to digest.
   a. The digestion may be halted when the quantity of residual fragments of adipose tissue is approx 5% of the initial amount.
6. On completion of digestion transfer the digestate to a fresh sterile glass separatory funnel. Allow the solution to sit for 5–10 min for phase separation to occur. Undigested and partially-digested adipose tissue, free adipocytes, and free lipid will float.
   a. The speed of phase separation may be increased by adding additional warm, sterile buffered saline to the funnel.
7. Open the stopcock and transfer the nonbuoyant fraction through a sterile 265 μm filter and into a sterile beaker.
8. Add warm, buffered sterile saline to the separatory funnel and invert the funnel 4–5 times with the cap in place. Return to the upright position and allow 3–5 min for phase separation.
9. Open the stopcock and transfer the nonbuoyant fraction through a sterile 265 μm filter into the material collected in step 7.
10. Aliquot the nonbuoyant solution collected in the beaker into multiple 50-mL centrifuge tubes.
11. Centrifuge at 400 *g* for 5 min at room temperature with a low-medium brake speed.

12. Gently pour off or aspirate the supernatant (top layer) into a liquid pathological waste container without disturbing the cell pellet.
13. Resuspend the pellets in buffered saline and combine the pelleted cells.
14. Repeat wash/centrifugation twice more to remove residual collagenase
15. Pass the cell suspension through a 100 μm cell strainer and collect into a new, sterile 50-mL centrifuge tube.
16. Perform cell counting using fluorescent live/dead dyes such as 7 amino-actinomycin D or propidium iodide in combination with a nuclear counterstain such as Acridine Orange *(25,26)* or systems that use esterase substrates *(27)*. Simple vital dye exclusion systems (for example Trypan Blue) that do not detect cell activity or the presence of a nucleus can be confounded by residual small lipid droplets.

In general this procedure yields a heterogeneous mixture of vascular cells, preadipocytes, lymphoid cells, blood cells, and ADSC. The process typically yields 2×10^4 nucleated cells per milliliter of human adipose tissue processed.

3.3 Assay of CFU-F

The CFU-F assay measures the presence of cells within the population capable of clonal expansion over 2 wk. Immunohistochemical staining has shown that these colonies are composed of cells that express CD105.

1. Centrifuge cells as above and resuspend in complete medium.
2. Plate cells in triplicate in 6 well plates at both 100 cells/cm^2 and 1,000 cells/cm^2.
3. Culture at 37 °C in humidified atmosphere of air plus 5% CO_2 for approx 2 wk.
   a. Perform scheduled media changes every 3–4 d.
4. After approx 2 wk of incubation, remove plates from the incubator and aspirate all medium from all wells.
5. Rinse each well of the plate 2–3 times with saline and then fix cells by incubating with ~1 mL of 10% neutral buffered formalin for 20 min.
6. Aspirate the formalin from each well and rinse the wells with saline.
7. Stain colonies by incubating wells with ~1 mL of hematoxylin Gill III formula for 5–10 min at room temperature.
8. Aspirate the stain and rinse gently with tap water.
9. Remove excess water by inverting and patting the plate onto a paper towel to dry the plate(s) and count the number of purple-stained colonies in each well consisting of more than 50 cells within a week of staining using an inverted microscope.

In our experience, application of the method described herein with freshly harvested human adipose tissue generates a population in which CFU-F represent approx 0.1–5% of nucleated cells.

4 Notes

1. ***Effect of body mass index on cell processing.*** In general we have found that overweight persons (persons whose body mass index (BMI) is between 27 and 30 kg/m^2 yield fewer nonbuoyant nucleated cells per unit volume of tissue than persons of lower BMI. This is likely owing to adipocyte hypertrophy (increased adipocyte size) in overweight persons such that there is more lipid and fewer cells per unit volume of tissue. Yield from tissue of obese persons is generally lower still. Data from one cohort of 23 donors (19 females, 4 males; median age 45; range 24–72 yr) is shown in Table 4.1.

   Body Mass index also affects the yield of CFU-F such that overweight persons yield significantly fewer clonogenic cells than persons of normal BMI (normal BMI $5.3 \pm 0.8 \times 10^3$ colonies/mL; overweight BMI $1.4 \pm 1.1 \times 10^3$ colonies/mL; $p = 0.012$). However, clonogenic cell yield from obese persons is highly variable perhaps as a result of the relative contribution of adipocyte hypertrophy and preadipocyte and stem cell hyperplasia in response to the increased demand for lipid storage in obese persons.
2. ***Rodent adipose tissue.*** Harvest of rodent adipose tissue is performed by lipectomy usually by dissection of the inguinal fat pad followed by mincing with scissors or scalpels. However, this tissue invariably contains lymph nodes which, if care is not taken to dissect them during mincing, will lead to considerable contamination of the nonbuoyant cell fraction with CD45-positive lymphoid cells and dilution thereby of the ADSC population. In one study digestion of murine inguinal adipose tissue without removing lymph nodes yielded a population in which CD45-positive cells comprised 94% of all nucleated cells and CFU-F frequency was 0.1%. By contrast, tissue in which major lymph nodes were dissected out before digestion yielded a population in which CD45-positive cells comprises 31% of all nucleated cells and CFU-F frequency was 4.7%.
3. ***Porcine adipose tissue.*** Contrary to the popular conception, pigs, especially juvenile farm swine bred for research purposes, are not particularly fat. Further, the tissue of these animals tends to have greater connective tissue than human fat rendering it less amenable to liposuction and to digestion. Hence, working with porcine adipose tissue requires more extensive digestion (0.4 U enzyme/mL of tissue, a prolonged digestion time (30–40 min), and care to ensure adequate mixing of tissue during digestion. In our hands porcine adipose tissue yields approx $5–10 \times 10^6$ nucleated cells/mL of tissue and a CFU-F frequency of 0.05–0.10%).
4. ***Sterility.*** Though liposuction is performed in a sterile surgical field, the containers into which lipoaspirate is collected during cosmetic liposuction are not usually sterile. As a result, testing of the crude lipoaspirate and the digestate sometimes shows the presence of bacterial contaminants. In our experience the washing and digestion processes, if performed properly, tend to reduce the content of bacteria as evidenced by the rate of positive bacterial cultures. Further, the presence of standard antibiotics in complete tissue culture medium is usually sufficient to avoid bacterial outgrowth and loss of cultures.

Table 4.1 Effect of body mass index on cell yield

Body mass index	Cell yield (million nucleated cells/mL tissue)
Normal (<27)	2.9 ± 0.4
Overweight (27–30)	2.2 ± 0.7
Obese (>30)	1.8 ± 0.7

References

1. Zuk, P. A., Zhu, M., Mizuno, H., Huang, J., Futrell, J. W., Katz, A. J., Benhaim, P., Lorenz, H. P., and Hedrick, M. H. (2001) Multilineage cells from human adipose tissue: implications for cell- based therapies. *Tissue Eng.* 7, 211–228.
2. Zuk, P. A., Zhu, M., Ashjian, P., De Ugarte, D. A., Huang, J. I., Mizuno, H., Alfonso, Z. C., Fraser, J. K., Benhaim, P., and Hedrick, M. H. (2002) Human adipose tissue is a source of multipotent stem cells. *Mol. Biol. Cell.* 13, 4279–4295.
3. Guilak, F., Lott, K. E., Awad, H. A., Cao, Q., Hicok, K. C., Fermor, B., and Gimble, J. M. (2006) Clonal analysis of the differentiation potential of human adipose-derived adult stem cells. *J. Cell. Physiol.* 206, 229–237.
4. Case, J., Horvath, T. L., Howell, J. C., Yoder, M. C., March, K. L., and Srour, E. F. (2005). Clonal multilineage differentiation of murine common pluripotent stem cells isolated from skeletal muscle and adipose stromal cells. *Ann. N Y Acad. Sci.* 1044, 183–200.
5. Hicok, K. C., Du Laney, T. V., Zhou, Y. S., Halvorsen, Y. D., Hitt, D. C., Cooper, L. F., and Gimble, J. M. (2004) Human adipose-derived adult stem cells produce osteoid in vivo. *Tissue Eng.* 10, 371–380.
6. Gronthos, S., Franklin, D. M., Leddy, H. A., Robey, P. G., Storms, R. W. and Gimble, J. M. (2001) Surface protein characterization of human adipose tissue-derived stromal cells. *J. Cell Physiol.* 189, 54–63.
7. Rodriguez, A. M., Pisani, D., Dechesne, C. A., Turc-Carel, C., Kurzenne, J. Y., Wdziekonski, B., Villageois, A., Bagnis, C., Breittmayer, J. P., Groux, H., Ailhaud, G. and Dani, C. (2005) Transplantation of a multipotent cell population from human adipose tissue induces dystrophin expression in the immunocompetent mdx mouse. *J. Exp. Med.* 201, 1397–1405.
8. Rodriguez, A. M., Elabd, C., Amri, E. Z., Ailhaud, G. and Dani, C. (2005). The human adipose tissue is a source of multipotent stem cells. *Biochimie.* 87, 125–128.
9. De Ugarte, D. A., Alfonso, Z., Zuk, P. A., Elbarbary, A., Zhu, M., Ashjian, P., Benhaim, P., Hedrick, M. H., and Fraser, J. K. (2003) Differential expression of stem cell mobilization-associated molecules on multi-lineage cells from adipose tissue and bone marrow. *Immunol. Lett.* 89, 267–270.
10. Winter, A., Breit, S., Parsch, D., Benz, K., Steck, E., Hauner, H., Weber, R. M., Ewerbeck, V., and Richter, W. (2003). Cartilage-like gene expression in differentiated human stem cell spheroids: a comparison of bone marrow-derived and adipose tissue-derived stromal cells. *Arthritis Rheum.* 48, 418–429.
11. Im, G. I., Shin, Y. W., and Lee, K. B. (2005). Do adipose tissue-derived mesenchymal stem cells have the same osteogenic and chondrogenic potential as bone marrow-derived cells? *Osteoarthritis Cartilage.* 13, 845–853.
12. Wagner, W., Wein, F., Seckinger, A., Frankhauser, M., Wirkner, U., Krause, U., Blake, J., Schwager, C., Eckstein, V., Ansorge, W., and Ho, A. D. (2005). Comparative characteristics of mesenchymal stem cells from human bone marrow, adipose tissue, and umbilical cord blood. *Exp. Hematol.* 33, 1402–1416.
13. Castro-Malaspina, H., Ebell, W., and Wang, S. (1984). Human bone marrow fibroblast colony-forming units (CFU-F). *Prog. Clin. Biol. Res.* 154, 209–236.
14. D'Ippolito, G., Schiller, P. C., Ricordi, C., Roos, B. A., and Howard, G. A. (1999). Age-related osteogenic potential of mesenchymal stromal stem cells from human vertebral bone marrow. *J. Bone Miner. Res.* 14, 1115–1122.
15. Stenderup, K., Justesen, J., Eriksen, E. F., Rattan, S. I., and Kassem, M. (2001). Number and proliferative capacity of osteogenic stem cells are maintained during aging and in patients with osteoporosis. *J. Bone Miner. Res.* 16, 1120–1129.
16. Nishida, S., Endo, N., Yamagiwa, H., Tanizawa, T., and Takahashi, H. E. (1999). Number of osteoprogenitor cells in human bone marrow markedly decreases after skeletal maturation. *J. Bone Miner. Metab.* 17, 171–177.

17. Muschler, G. F., Nitto, H., Boehm, C. A., and Easley, K. A. (2001). Age- and gender-related changes in the cellularity of human bone marrow and the prevalence of osteoblastic progenitors. *J. Orthop. Res.* 19, 117–125.
18. Mitchell, J. B., McIntosh, K., Zvonic, S., Garrett, S., Floyd, Z. E., Kloster, A., Halvorsen, Y. D., Storms, R. W., Goh, B., Kilroy, G., Wu, X., and Gimble, J. M. (2005). The immunophenotype of human adipose derived cells: Temporal changes in stromal- and stem cell-associated markers. *Stem Cells* 24, 376–385.
19. Fraser, J. K., Schreiber, R., Strem, B. M., Zhu, M., Wulur, I., and Hedrick, M. H. (2006). Plasticity of Human Adipose Stem Cells Towards Endothelial Cells and Cardiomyocytes: Therapeutic Perspectives. *Nature Clin. Pract.: Cardiovasc. Med* 3, S33–S37.
20. American Society for Aesthetic Plastic Surgery (2005). 2004 Cosmetic Surgery Quick Facts http://www.surgery.org/press/statistics.php.
21. Hunstad, J. P. and Aitken, M. E. (2006). Liposuction and tumescent surgery. *Clin Plast. Surg.* 33, 39–46.
22. Hunstad, J. P. and Aitken, M. E. (2006). Liposuction: techniques and guidelines. *Clin. Plast. Surg.* 33, 13–25.
23. Williams, S. K., McKenney, S., and Jarrell, B. E. (1995). Collagenase lot selection and purification for adipose tissue digestion. *Cell Transplant.* 4, 281–289.
24. Aust, L., Devlin, B., Foster, S. J., Halvorsen, Y. D., Hicok, K., du Laney, T., Sen, A., Willingmyre, G. D., and Gimble, J. M. (2004). Yield of human adipose-derived adult stem cells from liposuction aspirates. *Cytotherapy.* 6, 7–14.
25. Hathaway, W. E., Newby, L. A., and Githens, J. H. (1964). The acridine orange viability test applied to bone marrow cells. I. correlation with trypan blue and eosin dye exclusion and tissue culture transformation. *Blood.* 23, 517–525.
26. Laroche, V., McKenna, D. H., Moroff, G., Schierman, T., Kadidlo, D., and McCullough, J. (2005). Cell loss and recovery in umbilical cord blood processing: a comparison of postthaw and postwash samples. *Transfusion.* 45, 1909–1916.
27. Mueller, H., Kassack, M. U. and Wiese, M. (2004). Comparison of the usefulness of the MTT, ATP, and calcein assays to predict the potency of cytotoxic agents in various human cancer cell lines. *J. Biomol. Screen* 9, 506–515.

Chapter 5
Isolation of Human Adipose-derived Stem Cells from Biopsies and Liposuction Specimens

Severine G. Dubois, Elizabeth Z. Floyd, Sanjin Zvonic, Gail Kilroy, Xiying Wu, Stacy Carling, Yuan Di C. Halvorsen, Eric Ravussin, and Jeffrey M. Gimble

Abstract Adipose tissue has proven to serve as an abundant, accessible, and rich source of adult stem cells with multipotent properties suitable for tissue engineering and regenerative medical applications. Here, we describe a detailed method for the isolation and expansion of adipose-derived stem cells (ASCs). We present a large scale procedure suitable for processing >100 mL volumes of lipoaspirate tissue specimens and a small scale procedure suitable for processing adipose tissue biopsy specimens of <0.5 g. Although we have focused on the isolation of ASCs from human adipose tissue, the procedure can be applied to adipose tissues from other species with minimal modifications.

Keywords Adipose-derived stem cells (ASCs); biopsy; expansion; human; isolation; lipoaspirate; mesenchymal stem cells (MSCs);stromal vascular fraction (SVF).

1 Introduction

Adipocytes develop from mesenchymal cells via a complex cascade of transcriptional and nontranscriptional events that occurs throughout human life. Stromal cells that have preadipocytes characteristics can be isolated from adipose tissue of adult subjects, propagated in vitro and induced to differentiate into adipocytes *(1–4)*. Adipocyte differentiation is a complex process accompanied by coordinated changes in cell morphology, hormone sensitivity, and gene expression that have been studied primarily in murine preadipocyte cell lines rather than in human preadipocytes. This protocol describes primary in-vitro cultures of stromal cell isolated from either large or small quantities of human adipose tissue. This is of technical importance to minimize the amount of human tissue required to establish primary cultures. Furthermore, it will allow investigators to determine rates of proliferation and differentiation from limited amounts of human adipose tissue, such as those obtained by needle biopsy. Although historically in the literature adipose-derived stromal cells have been termed "preadipocytes" *(2,3)* there is a growing appreciation that they are multipotent, with chondrogenic, neurogenic, and osteogenic capability *(5–7)*. Consequently,

From: *Methods in Molecular Biology, vol. 449, Mesenchymal Stem Cells: Methods and Protocols*
Edited by D.J. Prockop, D.G. Phinney, and B.A. Bunnell

they have now been identified as adipose-derived stem cells or ASCs *(8)*. The ASCs have been distinguished from the plastic adherent adult stem/progenitor cells from bone marrow originally referred to as fibroblastoid colony forming units, then in the hematological literature as marrow stromal, subsequently as mesenchymal stem cells, and most recently as multipotent mesenchymal stromal cells (MSCs).

2 Materials

2.1 Tissue

1. Adipose tissue samples obtained from liposuction aspirates and needle biopsy.

2.2 Supplies

1. 200-mL plastic centrifugation bottles (Nalgene).
2. 0.2-μm filter units.
3. 50-mL conical tubes.
4. 2-mL tubes.
5. Scissors.
6. Hemocytometer.
7. Freezing apparatus (alcohol container).

2.3 Equipment

1. Inverted microscope, Nikon Eclipse TS100 with Epi-Fluorescence Attachment (Mercury Lamp Illuminator model name: C-SHG) (Nikon Instruments Incorporation, Melville, NY) and equipped with a camera photometric cool-snap (Nikon).
2. MetaMorph Imaging Software (Universal Imaging Corporation).
3. Shaking water bath.
4. Centrifuge.
5. Biosafety Hood.
6. CO_2 Incubator.

2.4 Media STOCK SOLUTION (see Note 1)

All the media solutions are filtered through a 0.2-μm filter unit.

1. *Collagenase solution:* Weigh out 0.1 g of type I collagenase and 1 g of powdered bovine serum albumin (fraction V). Dissolve these in 100 mL of phosphate buffered saline (PBS). After sterile filtration, warm the solution to 37 °C. This solution should be used with 1 h of its preparation.
2. *Stromal medium:* To 500 mL of DMEM/Ham's F-12 medium, add 55 mL of fetal bovine serum (10%) and 5.6 mL of antibiotic (penicillin/streptomycin)/antimy-

cotic (amphotericin) 100× stock solution. This solution should be used within 4 wk of its preparation. All fetal bovine serum should be prescreened before purchase for its ability to support both cell proliferation and adipocyte differentiation.

3. *Differentiation medium:* In advance, prepare and aliquot the following stock solutions and store frozen at −20 °C until required. (1) A 66 m*M* stock solution of biotin (2,000-fold concentration) dissolved in 1 *N* sodium hydroxide. (2) A 34 m*M* stock solution D-pantothenate (2,000-fold concentration) dissolved in water. (3) A 1 m*M* dexamethasone (1,000-fold concentration) dissolved in water or ethanol depending on its formulation. (4) A 250 m*M* stock solution of methyl-isobutylxanthine (1,000-fold concentration) dissolved in dimethyl sulfoxide. (5) A 200 μ*M* stock solution of human insulin (2,000-fold concentration) dissolved in PBS. (6) A 5 m*M* stock solution of rosiglitazone or equivalent PPARγ agonist dissolved in dimethyl sulfoxide. Use this solution within 2 wk of its preparation.
4. *Adipocyte maintenance medium:* This solution is prepared in an identical manner as differentiation medium except that it does not contain either the isobutyl-methylxanthine or the PPARγ agonist; these 2 stock solutions should be omitted. Use this solution within 2 wk of its preparation.
5. *Freezing medium:* The freezing medium consists of 80% fetal bovine serum, 10% DMEM/Ham's F-12, and 10% dimethyl sulfoxide. Use this solution within 2 wk of its preparation.
6. *Medium Sterility Test Procedure:* Before use, it is wise to test the sterility of the medium by removing a single milliliter from each bottle, placing it in a single well of a 24-well plate, and incubating it for 48 h in a humidified, 37 °C, CO_2 incubator. After this period, examine the plate using a phase contrast microscope for any evidence of contamination. If contaminated, immediately inactivate all bottles and test plates with 15% bleach solution and discard.

3 Methods

After transportation to the laboratory, the liposuction sample can be kept at room temperature for no more than 24 h before use. Before performing the experiment, warm up the water bath to 37 °C.

All the following procedures are performed in biosafety hoods. Investigators should be trained in the handling of human tissues and human pathogens before initiation of any studies.

3.1 Small Volumes of Adipose Tissue from Needle Biopsy

1. Warm up buffer (PBS + 1% antibiotic).
2. Warm up freshly prepared *collagenase solution* in the 37 °C water bath.
3. Prepare PBS (or KRB) solution with 1% BSA, filter the solution and warm it in the 37 °C water bath.
4. Prepare stromal medium: *cf Media stock solution*. This should have been done in advance of the procedure.

5. Tissue washes:
   A small amount of tissue (up to 150 mg) is placed in a 2-mL tube containing 500 µl of warm PBS with 1% antibiotic solution. The sample is then washed with 300 µl of warm PBS containing 1% antibiotic solution. The washing step is repeated until all blood vessels and connective tissues appear to have been liberated (usually 2 washes). To facilitate the subsequent tissue digestion, mince the adipose tissue sample into small pieces using sterilized scissors.

6. Tissue digestion:
   If the volume of adipose tissue sample is small (150–250 mg), it is recommended that you increase the volume of collagenase solution (up to 400 µl for an equivalent of 150–250 mg) to improve the efficiency of the tissue digestion. Wrap the tube(s) with parafilm and place into the 37 °C shaking water bath at ~75 rpm for 60 min until the tissue appears smooth on visual inspection (see Note 2). Because over-digestion can damage the cells and reduce the final yield, you should inspect the reaction visually during the incubation and stop the reaction, as soon as the tissue appears smooth and fully digested.

7. Isolation stromal vascular fraction (SVF):
   After digestion, spin the samples at 300 g in an appropriate centrifuge for 5 min at room temperature. Take the samples out of the centrifuge and shake them vigorously to thoroughly disrupt the pellet and to mix the cells. This is to complete the separation of the stromal cells from the primary adipocytes.

Repeat the centrifugation step.

8. Pellet clean-up:
   After spinning, aspirate all the collagenase solution above the pellet without disturbing the cells. Add to each tube a volume of 200 µl of warm PBS solution containing 1% BSA. Centrifuge the cells at 300 g in an appropriate centrifuge for 5 minutes at room temperature.

   Re-suspend the cells with 200 µl of stromal medium. Centrifuge the cells at 300 g in an appropriate centrifuge for 5 min at room temperature.

9. Plate the cells:
   After spinning, aspirate the supernatant and suspend the cells in 100 µl of stromal medium (*see* **Note 3**).

   Innoculate the cells in a single well of a 12-well plate for an amount of about 500 mg of adipose tissue or in a single well of a 24-well plate for an amount of 250–150 mg of adipose tissue (*see* **Note 4**). Then add a volume of stromal medium according to the well capacity of the culture plate (see Table 5.1).

Table 5.1 Table for plating

Plate	Area per plate	Cells per plate	Cells per well	Media per well
6-well plate	60 cm^2	1.8×10^6	30×10^4	2.5 mL
24-well plate	48 cm^2	1.44×10^6	6×10^4	1 mL
96-well plate	31 cm^2	0.93×10^6	10^4	200 µL

The samples are then maintained in a humidified incubator at 37 °C with 5% CO_2.

10. Change stromal medium:
Seventy-two hours after plating, aspirate the entire medium from the wells (*see* **Note 5**). Wash the cells with prewarmed PBS (1% antibiotic can be added to the solution) by pipeting up-and-down to clean the cells thoroughly from any tissue fragments and/or blood cells. Add a volume of fresh stromal medium according to the well capacity of the culture plate.

Critical steps

The medium is then changed every 2–3 d until the cells achieve 80–90% confluence.
⇒ Troubleshooting

11. Harvesting cells:
When the cells reach 80–90% confluence there are 2 options: either harvest the cells or directly induce the adipocyte differentiation (*see* Step 12). To harvest the cells, the following procedure can be used:
Remove medium from wells and save the sterile "conditioned media" in a sterile tube for future cell culture application (this should be sterile filtered before use). Add a small volume (250–500 µl) of sterile warm PBS to the wells and allow PBS to remain on cells for 2 min. Replace the PBS with 500 µl of Trypsin/EDTA solution (0.5%). Incubate in incubator for 5–10 min. Verify under microscope that more than 90% of the cells have detached and then add 500 µl of stromal medium to allow the serum contained in the solution to neutralize the trypsin reaction.

Critical steps

Transfer the medium containing the suspended cells from the well to a sterile 2-mL tube. Centrifuge at 300 g for 5 min. Aspirate the supernatant and suspend the cells in a small volume of stromal medium (~250 µl).
Proceed to cell counting by taking an aliquot of cells diluted in trypan blue (for a 1:8 dilution: add 12.5 µl of suspended cells to 87.5 µl of trypan blue). Count cells using the hematocytometer.
After counting, cells can then be replated according to the well capacity in adequate cell culture plates.

12. Adipocyte differentiation:
When the cells reach between 80 and 90% confluence (before or after harvesting the cells), the preadipocytes are induced to differentiate. Aspirate the medium, add a small volume (about 1.5 mL for a 6-well plate) of prewarmed PBS + 1% antibiotic to wash the cells, and then remove the PBS by aspiration (*see* **Note 6**). Next, add the differentiation medium.
The cells will be maintained in the differentiation medium for 3 d.

13. Day +3 differentiation:
    Aspirate the differentiation medium and wash the cells with prewarmed PBS + 1% antibiotic (*see* **Note 7**).
    Then add a volume (2.5–3 mL for a 6 well plate) of adipocyte medium.
    The adipocyte medium will be changed every 3 d until mature adipocytes are obtained (Day +9-+12 differentiation) (*see* **Note 8**).
14. Fixation of cells:
    After 12 d of differentiation, the cells can be fixed either using a 10% formalin solution or 70% ethanol[a]. After removing the medium and washing the cells with PBS, immerse the cells in the fixative solution: 10% formalin or 70% ethanol for 30 min or 1 h respectively. Remove the fixative and air dry before staining or add water until performing the dye-staining (fixed cells can be stored at 4 °C or −20 °C for as long as several months)
    [a]Using ethanol, there is a risk that the lipids will be eluted from the cells.
15. Cell staining:
    This step does not need to be performed under a biosafety hood. To distinguish the lipid filled cells (mature adipocytes) and nondifferentiated cells (preadipocytes), a 2-step dye-staining is performed by using the neutral lipid fluorescent dye (BODIPY)[1] and the nuclear fluorescent dye (DAPI)[2] (*see* **Note 9**). All the procedures are performed at room temperature.

    [1]The maximum fluorescence emission range is from 510 to 665 nm.
    [2]The maximum excitation of DAPI is in the near-ultraviolet (UV) range at about 360 nm and the emission maximum in the blue at 460 nm.

    a. To prepare the BODIPY working solution, dilute the BODIPY stock solution (1 mg/mL ethanol) to 10 μg/mL in H_2O (at this concentration, the background signal could be high; in that case, use a lower concentration (1 μg/mL) to minimize it).
    b. Prepare DAPI working solution: dilute DAPI stock solution (5 mg/mL H_2O) to 300 n*M* in PBS ⇒ (*see* **Note 9**) Troubleshooting
    c. **BODIPY staining** - Immediately after drying the cells from the fixative solution or after removing the water from the wells, add BODIPY working solution and stain for 20 min (it is recommended to perform the staining in a semi-dark environment).
       After 20 min, remove all BODIPY solution and immediately wash with H_2O several times.
    d. **DAPI staining:** Remove all H_2O from the wells and air dry the cells for 1–2 min. Add DAPI working solution to the cells and counterstain for 20 min at room temperature.
    e. Percentage of Differentiation (PDIFF): The percentage of cells undergoing adipogenesis can be calculated by microscopic inspection. The number of cells in a field staining positive with BODIPY for lipid droplets can be determined as a percentage relative to the total number of cells in the field, i.e., the number of cells determined by positive staining with the DAPI nuclear stain. This can be performed using a fluorescent microscope and

appropriate image capturing software such as a Nikon Eclipse TS100 equipped with an epi-fluorescence attachment (Mercury Lamp Illuminator), a photometric cool-snap camera (Nikon Instruments, Melville, NY), and Metamorph software (UIC, Downingtown, PA).

3.2 Large Volumes of Tissue (Liposuction Aspirates ≥ 100 mL)

1. Warm up buffer (500 mL or more of PBS or KRB). Line the surface of the biosafety hood with a disposable bench protector.

2–4. Identical to Subsection 3.1 Steps 2–4.

5. To maintain optimal sterile conditions, open the surgical container used for liposuction procedure under the biosafety hood (*see* **Note 10**).

Critical steps

Dispense a volume of adipose tissue in sterile plastic bottles (Nalgene): for each 225 cm^2 Flasks (0.16 mL tissue/cm^2) it is recommended that you distribute about 35 mL of tissue; each bottle can accommodate ~70 mL of tissue. We routinely process a total of 140 mL of tissue to be plated in four 225 cm^2 flasks. Add an equal volume of warm PBS. Agitate to wash the tissue and then allow phase separation for 3–5 min. Suction off the infranatant solution (lower liquid phase). The wash is repeated several times until a clear infranatant solution is obtained (usually 3–4 times).

6. Add an equal volume (60–70 mL) of warm collagenase solution into the Nalgene bottles containing the clean adipose tissue sample. Wrap the bottle tops in parafilm and place the bottles in a 37 °C shaking water bath at ~75 rpm for 60 min until the tissue appears smooth on visual inspection.
7. Identical to Subsection 3.1. Step 7.
8. After spinning, the stromal vascular fraction will form a pellet at the bottom of the bottle or tube (this will usually include a layer of dark red cells). Carefully remove the top layer of oil and fat, the primary adipocytes (a yellow layer of floating cells), and the underlying layer of collagenase solution. Leave behind a small volume of collagenase solution above the pellet so that the cells are not disturbed.

   Resuspend the cells in 10 mL of warm PBS (or KRB) solution containing 1% BSA and transfer the solution containing the cells into a 50-mL conical tube. Centrifuge the cells at 300 X g in an appropriate centrifuge for 5 min at room temperature.

   Aspirate the remaining collagenase solution. When aspirating, the tip of the pipet should aspirate from the top so that the oil is removed as thoroughly as possible. The cell pellet should be at the bottom of the tubes. Resuspend the cells with 10 mL of stromal medium in each tube. Pool the cells in one 50-mL conical tube and spin the cells at 300 X g in an appropriate centrifuge for 5 min at room temperature.
9. After spinning the cells, aspirate off the supernatant and resuspend the cells in 10 mL of stromal medium.

Divide the cells according to the number of flasks. The cells are plated at a density equivalent to approx 0.16 mL of liposuction tissue aspirate per cm^2 of surface area (volume of ~35 mL of tissue for a 225-cm^2 flask).

Divide the cells according to the number of flasks. In this protocol, we use about 140 mL of liposuction tissue. Thus, to each 225 cm2 flask (×4), we add 2.5 mL of cell suspension and 37.5 mL of stromal medium.

10. At 48 h[a] after plating, aspirate the medium from the flask. Wash the cells with prewarmed PBS (*see* **Note 11**). Add 40 mL of fresh stromal medium.

    Critical steps

    [a]This period can vary from 24 to 72 h depending on the number of cells attached to the plastic surface (observed under microscope).

    The medium is then changed every 2–3 days until the cells achieve 80-90% confluence.

    ⇒ Troubleshooting

11. Harvesting cells:
    Remove medium from flasks and save the sterile “conditioned media” in a sterile tube for future cell culture application (this should be sterile filtered before such use). Add 10 mL of sterile warm PBS to the flasks and allow PBS to remain on cells for 2 min while flasks are in a horizontal position. Replace the PBS with 10 mL of Trypsin/EDTA solution (0.5%) (*see* **Note 12**). Incubate in incubator for 5–10 min. Verify under microscope that more than 90% of the cells have detached and then add 10 mL of stromal medium to allow the serum contained in the solution to neutralize the trypsin reaction.

    Critical steps

    Transfer the medium containing the suspended cells from the flask to a sterile 50-mL conical tube. Centrifuge at 300 X g for 5 min. Aspirate the supernatant and suspend the cells with a small volume of stromal medium (about 2 mL).

    Proceed to cell counting by taking an aliquot of cells diluted in trypan blue (for a 1:4 dilution: add 25 µl of suspended cells to 75 µl of trypan blue). Count cells using the hematocytometer.

12. After counting, you have several options:

    a. Cryopreservation:
       Suspend the cell pellet in room temperature freezing medium at a concentration of 2×10^6 cells/mL. Dispense 1 mL of aliquots of the cell suspension to sterile cryovials. Place cryovials in appropriate freezing apparatus (alcohol container). Freeze cells to –80 °C. The next day, transfer the cells on dry ice or other frozen material to a liquid nitrogen storage container.
    b. Use of cells for protein extraction or flow cytometry (not described in this protocol)
    c. Re-plating the cells:

After cell count, suspend the cell pellet in stromal medium following the different concentrations listed in Table 5-1 to achieve a confluent culture within 24 h of re-plating.

4 Notes

1. Before purchase, the fetal bovine serum should be assayed to test for its ability to support adipogenesis. DAPI is a mutagen reagent and it should be handled with care.
2. **After 1 h digestion, some pieces of undigested tissue are seen in the tube.** Make sure that the collagenase solution is fresh and has not been maintained at room temperature for an extended period of time. This is necessary to maximize the enzyme efficiency. The collagenase solution can also be stored at –20 °C for a few days, with a minor loss of enzyme activity. Before use, the frozen solution is slowly thawed at room temperature and prewarmed to 37 °C. However, it is all right to not complete the digestion if a small amount of tissue fragments are observed in the solution.
3. A filtration procedure can be performed by using a nylon mesh filter with a small pore size (250–350 μm). The suspension is then centrifuged at 300 g at room temperature to allow separation of the stromal vascular fraction from the mature adipocytes. The filtration can be performed after removing the floating mature adipocytes cells. This step will remove tissue fragments and is used by some investigators. However, the filtration procedure is not recommended for small amounts of adipose tissue.
4. To accelerate cell adhesion, the culture dishes can be precoated with extracellular matrix proteins, such as gelatin or Matrigel.
5. **Cells do not grow very well and do not appear healthy.** The percent of preadipocytes obtained from the stromal vascular fraction after digestion is patient-dependent. Supplementation of the culture medium with 20–30% conditioned medium (saved from previous cultures) should facilitate the growth of the cells. If you do not have conditioned medium, another alternative is to increase the FBS contained in the stromal medium to 15–20%; however this may promote premature adipogenesis. Signs of deterioration such as granularity around the nucleus, cytoplasmic vacuolations, and/or detachment of the cells from the plastic surface may indicate inadequate or toxic medium, microbial contamination or senescence of the primary cells.
6. **Adipocytes are detaching.** Do not dry the well when changing the medium because adipocytes tend to float when new medium is added.
7. According to the protocol, the medium is changed every 2–3 d. However, as the adipocytes mature, you may observe a yellowing of the culture medium: a drop in pH may account for this. As the pH falls from 7 to 6.5, cell growth will decline and cell viability falls at pH between 6.5 and 6. You can observe this change of pH by looking at the medium color change, going from red (pH 7) through yellow (pH $\leq$ 6), indicating the need for an immediate change of the medium.
8. **Cells do not differentiate very well.** The differentiation process is patient dependent. The age of the donor can be a factor, because some studies suggest that the differentiation capacity is higher in culture from younger subjects compared to older people. To further enhance adipogenesis, the following alternatives are proposed:

   - You may try different PPARγ agonists (troglitazone, pioglitazone among others).
   - The addition of 5% rabbit serum (RS) can be added to the differentiation medium enhance differentiation (the ethyl acetate contain in the RS has been found 35-fold more abundant than in FBS *(9)*.
   - Another alternative would be to perform the addition of the differentiation medium multiple times after a three day rest period; i.e., 3 d on in the presence of the differentiation medium and three days off in the presence of the adipocyte medium. Repeat this cycle until mature adipocytes are obtained *(9)*. It takes some time to completely dissolve DAPI in water. The manufacturer (Molecular Probes) recommends that the sample be completely dissolved by sonication. The stock solution is kept at 2–6 °C, protected from light.
9. While stored, it is possible that DAPI crystals fall down of solution so it is recommended that you remix the stock solution properly before preparing the working solution. To keep a maximal efficiency of the dye fluorescent, it is recommended that you store BODIPY and DAPI in containers wrapped in aluminum foil and protected from light exposure.

10. **Contamination –Microbial**. To keep optimal sterile conditions, it is recommended that you open and close the container properly to avoid any potential contaminations. It is vital that the culture be examined regularly to confirm the absence of microbial contamination. To avoid this problem, 5% iodine solution can be added to the initial wash solution. The disinfectant is then washed away in subsequent washes. To avoid this problem in small adipose tissue samples, it is recommended to add in the PBS solution 1% of antibiotic solution and wash thoroughly the cells with this solution.
11. **Contamination-Cellular.** The blood cells could be a source of contamination and may reduce or prevent adhesion of stromal cells; it is then important to wash thoroughly the cells with PBS (1% of antibiotic solution can be added). When removing the PBS from the cells, aspirate the solution up-and-down until the cells appear clean and free of red cells. An erythrocyte lysis buffer (155 mM NH_4Cl, 5.7 mM K_2HPO_4, 0.1 mM EDTA et pH 7.3) can serve to remove red blood cells. Endothelial cells (EC) could be another source of contamination. Intra-abdominal depots are more subject to this type of contamination when compared to subcutaneous adipose tissue, which is essentially free of EC. Therefore a filtration procedure can be performed by using a nylon mesh filter with a small pore size (25 μm).
12. It is important to not overexpose the cells to the trypsin/EDTA solution. This could decrease the cell viability.

Final Comments

Although valuable preadipocyte cell lines of rodent origin have been available for more than 30 years, substantial differences exist between the development, regulation and function of rodent and human adipocytes. Recently, there has been renewed interest in primary cell culture models allowing the study of human adipocyte differentiation in vitro. This protocol is designed to provide new laboratories with the tools to cultivate human adipocyte precursor cells as they explore adipose tissue metabolism. Notably, the first part of this protocol is designed for small amounts of adipose tissue as samples obtained from needle biopsy and will give investigators the capability to obtain primary adipocytes cells from a variety of patients with metabolic diseases. With improved technologies, cellular components from biopsy samples could be use for microarray, proteomics, and many other applications. This approach will help to better understand the particular role of adipose tissue in metabolic diseases such as type-2 diabetes. The second part of this protocol describes the scaled-up version of this procedure for use with larger liposuction samples *(10)*.

References

1. Rangwala, S. M., and Lazar, M. A., (2000) Transcriptional control of adipogenesis. *Annu. Rev. Nutr*, **20**; 535.
2. Deslex, S., Negrel, R., Vannier C., Etienne J., and Ailhaud G., (1987) Differentiation of human adipocyte precursors in achemically defined serum-free mediu..*Int.J.Obes*, **11**: 19–27.
3. Hauner, H., *Entenmann, G., Wabitsch, M., Gaillard, D., Ailhaud, G., Negrel, R.*, and *Pfeiffer, E.F.* (1989) Promoting effect of glucocorticoids on the differentiation of human adipocyte precursor cells cultured in a chemically defined medium. *J. Clin. Invest.* **84**:1663–1670.
4. Halvorsen, Y. D., Bond, A., Sen, A., Franklin, D. M., Lea-Currie, Y. R., Sujkowski, D., Ellis, P. N., Wilkison, W. O., and Gimble, J. M. (2001) Thiazolidinediones and glucocorticoids synergistically induce differentiation of human adipose tissue stromal cells: biochemical, cellular, and molecular analysis. *Metabolism*. **50**:407–413.

5. Zuk, P. A., Zhu, M., Ashjian, P., De Ugarte, D. A., Huang, J. I., Mizuno, H., Alfonso, Z. C., Fraser, J. K., Benhaim, P., and Hedrick, M. H. (2002) Human adipose tissue is a source of multipotent stem cells. *Mol. Biol. Cell.* **13**:4279–4295.
6. Guilak, F., Lott, K. E.., Awad. H. A., Cao. Q., Hicok. K. C., Fermor. B., Gimble. J. M. (2006) Clonal analysis of the differentiation potential of human adipose-derived adult stem cells. *J. Cell. Physiol..* **206**:229–237.
7. Gimble, J. M., and Guilak, F. (2003) Differentiation potential of adipose derived adult stem (ADAS) *Cells.Curr. Top. Dev. Biol.* **58:**137–160.
8. Mitchell, J. B., McIntosh, K, Zvonic, S., Garrett, S., Floyd, Z. E., Kloster, A., Halvorsen, Y. D., Storms, R. W., Goh, B., Kilroy, G., Wu, X., and Gimble, J. M. (2006) The immunophenotype of human adipose derived cells: Temporal changes in stromal- and stem cell-associated markers. *Stem Cells*, **24:**376–385.
9. Diascro, D. D., Vogel, R. L., Johnson, T. E., Witherup, K. M., Pitzenberger, S. M., Rutledge, S. J., Prescott, D. J., Rodan, G. A., Schmidt, A. (1998) High fatty acid content in rabbit serum is responsible for the differentiation of osteoblasts into adipocyte-like cells, *J Bone Miner Res*, **13**:96–106.
10. Dubois, S. G., Halvorsen, Y. D. C., Ravussin, E., and Gimble, J. M. (2005) Primary stromal cell culture from adipose tissue: from liposuction to needle biopsy. *Adipocytes* **1:**139–144.

Part II
Manipulation of Human Multipotential Stromal Cells

Chapter 6
Colony Forming Unit Assays for MSCs

Radhika Pochampally

Abstract MSCs are the plastic adherent adult stem/progenitor cells from bone marrow originally referred to as fibroblastoid colony forming units, then in the hematological literature as marrow stromal cells, subsequently as mesenchymal stem cells, and most recently as multipotential mesenchymal stromal cells. MSCs were originally referred to as fibroblastoid colony-forming-cells because one of their characteristic features is adherence to tissue culture plastic and generation of colonies when plated at low densities *(1,2)*. The efficiency with which they form colonies still remains an important assay for the quality of cell preparations. This chapter describes 2 methods to assay the colony forming ability of MSCs; (a) a traditional assay for colony forming units (CFUs) and (b) single-cell colony forming unit assay (Sc-CFU).

Keywords Mesenchymal stem cells; marrow stroma cells; multipotential stromal cells; single-cell colony forming unit assay.

1 Introduction

Human MSCs proliferate rapidly and largely retain their multipotentiality for differentiation in culture. However, cultures of expanded cells are heterogeneous in morphology and in their content of the earliest progenitor cells. Also, the cultures gradually loose multipotentiality as they are replated for 6 or 7 passages *(3,4)*. The cells are highly sensitive to plating density, and early progenitors are rapidly lost if the cultures are grown to confluency *(3,5)*. Additionally, there is considerable variation in the proportion of early progenitors recovered from different samples of bone marrow, even when the samples are obtained from the same donor at the same time. Currently, there are no surface epitopes that are useful for distinguishing early progenitors from mature cells in the cultures *(6,7)*. For these reasons, it is obviously important to devise standardized assays for characterizing MSCs.

From: *Methods in Molecular Biology, vol. 449, Mesenchymal Stem Cells: Methods and Protocols*
Edited by D.J. Prockop, D.G. Phinney, and B.A. Bunnell © Humana Press, Totowa, NJ

As originally indicated by Friedenstein *(1)*, one of the most prominent properties of MSCs is their ability to generate colonies after they are plated at low density. In this chapter we describe what has become a traditional assay of MSCs for colony-forming units (CFU assay) in which the cells are plated at low density in large plates and discrete colonies counted after 2 or 3 wk. When used for assay of human MSCs, each colony is generated by a single cell. When used with rat or mouse MSCs, however, single cells can generate more than one colony because the cells can detach as they expand and re-seed the plate *(8,9)*. In this chapter we also describe a refined assay in which single MSCs are plated into individual wells of a microtiter plate with a protocol that ensures that each observed colony was generated by a single cell (Sc-CFU assay). With the Sc-CFU assay it is possible to distinguish the colony forming potential of 2 distinct kinds of MSCs present in early passage cultures: (a) spindle shaped cells that are rapidly selfreplicating (RS-cells) that predominant in the first few days after plating the cells at low density, and (b) broader, slowly replicating cells (SRcells) that predominate as colonies or cultures become confluent. The RS-cells can also be distinguished from SR-cells by their lower forward scatter (FS^{lo}) and lower side scatter (SS^{lo}) of light, but the assays of FS/SS are difficult to standardize. Therefore the CFU assays are more useful in estimating the proportion of early progenitors in different preparations of MSCs. To develop an improved assay for CFUs, we employed a fluorescent flow cytometer with an automated cell sorter (FACSVantage SE with Clonesort accessory; Becton-Dickinson) to plate single cells into individual wells of a 96-well microtiter plate. We then incubated the samples in complete medium for 10–14 d and assayed visible colonies by staining the plates with Crystal Violet. As indicated in Figure 1A, the single-cell CFU assay (sc-CFU) had a smaller variation than the standard CFU assay (Figure 1A and 1B). The average coefficient of variation *(10)* was 4.52 for the sc-CFU and 14.6 for the standard CFU assay. Therefore, the sc-CFU assay was about 3 times more reproducible. Also, the sc-CFU assay detected important differences not detected by the standard assay (Figure 1B) between cultures initially plated at 50 or 100 cells/cm^2 and cultures plated 500 or 1,000 cells/cm^2. The lower values obtained with the sc-CFU assay for cultures plated at the higher density are consistent with previous observations that cultures plated at higher density show a rapid decrease in the number of multipotential and rapidly self-renewing cells (RS cells) *(9,11–13)*.

2 Materials

2.1 Reagents

1. αMinimum Essential medium (αMEM) without deoxyribonucleotides (GIBCO/BRL).

2. Fetal bovine serum (Atlanta Biologicals)—selected for enhanced cell growth and low alkaline phosphatase (AP) activity.
3. Penicillin 10,000 units/mL/ 10,000 μg/mL Streptomycin (GIBCO/BRL – Cat# 15140-122).
4. L-glutamine (200 m*M*) (GIBCO/BRL).
5. Phosphate Buffered Saline (PBS), 1× (GIBCO/BRL).
6. 0.25% Trypsin in 1 mM ethylene diamine tetraacetic acid (EDTA) (Sigma).
7. 0.4% Trypan Blue (Sigma).
8. Crystal Violet (Sigma).
9. Methanol (100%).
10. Annexin V-FITC (Sigma).

2.2 Equipment and Supplies

1. 5-, 10-, and 25-mL sterile serological pipets.
2. Sterile transfer (Pasteur) pipets.
3. 50-mL sterile conical centrifuge tubes.
4. 15-mL sterile conical centrifuge tubes.
5. 10-cm (58 cm2) and/or 15-cm (152 cm2) sterile tissue culture dishes or T75 (75 cm2) flasks.
6. Laminar Flow Hood (Biosafety Cabinet).
7. 70% ethanol.
8. Water bath (37 °C).
9. Humid Incubator (37 °C) with CO2 source.
10. Vacuum aspiration source.
11. Hemocytometer.
12. Centrifuge with swinging bucket rotor.
13. Inverted Microscope (for checking culture confluence).

2.3 Cell culture

1. Complete Culture Medium (CCM) with 18% FBS: 500 mL αMEM with or without deoxyribonucleotides, 100 mL FBS, 5 mL 10,000 units/mL Penn/10,000 μg/mL Strep (Gibco/BRL, Cat# 15140-122) for a final of 100 units Pen/mL and 100 ug/mL Strep, 5 mL 200 m*M* L-glutamine, 29.2 mg/mL of 0.85% NaCl for a final concentration of 2 m*M* L-glutamine (292 μg/mL).
2. 3.0% Crystal Violet in 100% Methanol
   3.0 g Crystal Violet in 100 mL 100% Methanol.
   Filter through Whatman filter paper. Store at room temperature (RT)

3 Methods

3.1 Colony Forming Unit—Fibroblast assay

1. Human MSCs are prepared as described previously in Chapter 1.
2. Expand MSCs cultures to 70–80% confluency and harvest with trypsin-EDTA (as described previously).
3. Count the number of cells using a hemocytometer (*see* **Note 1**). To ensure cell separation, a glass Pasteur pipet can be flamed at its tip to reduce its diameter, and the cells drawn through the narrowed pipet several times (*see* **Note 2**).
4. Dilute in complete culture medium, and plate at about 100 cells per 100-mm tissue culture dish (Falcon) in complete culture medium.
5. Incubate for 10–14d at 37°C in 5% humidified CO2, and wash with PBS and stain with 0.5% Crystal Violet in methanol for 5–10min at room temperature.
6. Wash the plates with PBS twice and visible count colonies (Fig. 6.1).

3.2 Colony Forming Unit— *Single Cell assay*

Rapidly self-renewing population of MSCs are characterized by low forward scatter (FS^{lo}) and low side scatter (SS^{lo}) of light. The following experiments explain the isolation of FS^{lo}/SS^{lo} MSCs that are rapidly self-renewing. It is also a rapid, standardized assay for FS/SS useful to identify preparations of MSCs enriched for RS cells that will expand rapidly during subsequent passage in culture. The use of the assay should help to resolve discrepancies in data obtained by different laboratories with apparently similar preparations of MSCs.

3.2.1 Measurement of FS and SS

1. A closed stream flow cytometer (Epics XL/ADC; Beckman-Coulter) is standardized using microbeads with known diameters (7, 10, 15, and 20 microns; Dynosphere Uniform Microspheres; Bangs Laboratories Inc., Fisher, IN).
2. The gains and voltages on the photomultiplier tubes are adjusted so that the mean value of the FS peak for the 20µ bead is about 650 and the peak of the SS for the 7µ*M* bead about 450.
3. With these settings, the standard deviation for FS of the largest bead should be less than +/– 0.4% ($n = 3$) of the mean and the slope of FS on a linear scale of 0 to 1,023 at least 41 (*see* **Notes 3**) (Fig. 6.2).
4. For the assays, cells are lifted with trypsin/EDTA, washed with CCM by centrifugation at 450*g* for 10min to neutralize trypsin, counted on a hemocytometer,

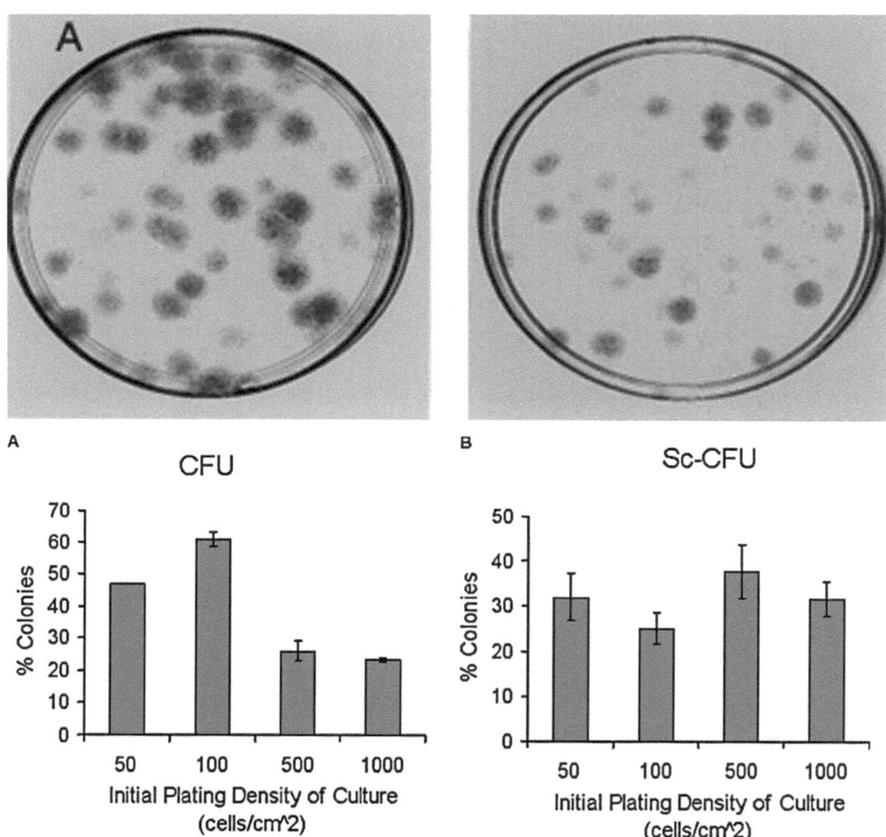

Fig. 6.1 (**A**) Crystal violet stained plates of CFU-F assays performed on two different donors *(10)*. (**B**) Sc-CFU is more sensitive and reproducible than the traditional CFU assay: A: Sc-CFU assay of MSCs initially plated at varying densities and incubated for 10–11 d (mean+/−SD, n=2). B: Standard CFU assay of MSCs plated at same densities and incubated for 13–14 d (mean+/-SD, *n* = 3 or 4) (*See Color Plates*)

suspended in PBS at 4 °C at a concentration of about 0.5 million cells/mL, and then assayed shortly thereafter (*see* **Notes 4**).

5. Staining with Annexin V-FITC demonstrates that the events in the upper left of the plot are cell debris and dead cells (R1 in Fig. 6.3B). To obtain sub-fractions of cells, the Annexin V^+ events are gated out and four sub-populations were defined on the basis of FS and SS (Fig. 6.3C).
6. For single cell sorting the cells treated with the Annexin V-FITC and maintained at 4 °C to prevent aggregation from the presence of calcium and reagent-induced toxicity.
7. To isolate distinct fractions on the basis of FS and SS (Fig. 6.3C), divided the Annexin V^- events into 4 quadrants on the basis of FS and SS.

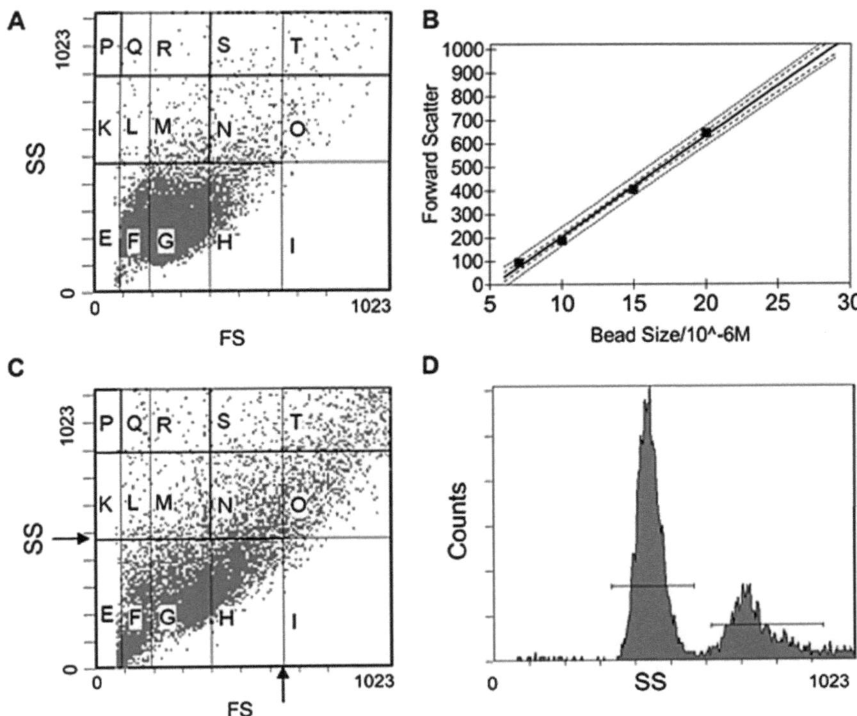

Fig. 6.2 Standardized assay for FS/SS of MSCs. **(A)** Assay of passage 3 MSCs from a culture initially plated at low density (100 cells per cm²) and incubated for 4 d. The cells were lifted with trypsin/EDTA and then assayed on the closed stream flow cytometer (see Methods). **(B)** Calibration curve obtained with microbeads to standardize measurement of FS. The calibration curve was used to locate the vertical lines in **A** and **C.** **(C)** Assay of passage 5 MSCs from a culture initially plated at high density (1,000 cells per cm²) and incubated for 4 d. Arrows indicate FS for 20-µ*M* bead SS set at 650 and SS for the 10-µ*M* bead set at 450. **(D)** SS calibration indicating the two peaks in SS corresponding to the 7- and 10-µ*M* scatter; beads that were used to place the horizontal lines in **(A, C)**.Abbreviations: FS, forward scatter; MSC, marrow stromal cell; SS, side scatter (Figure reproduced with permission from Stem Cells, 2005, 22(5))

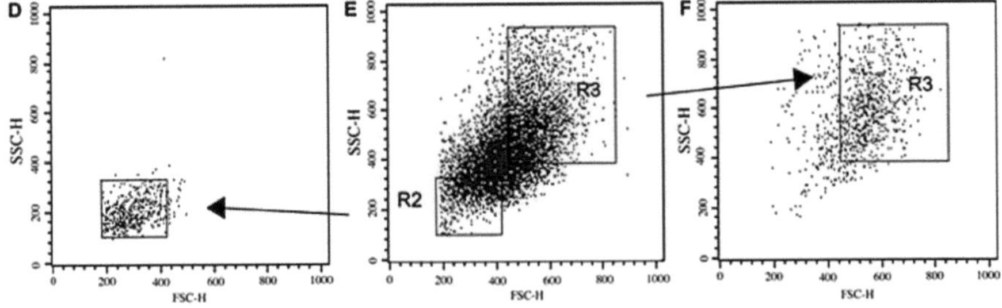

Fig. 6.3 Representative assay of FS/SS of passage 2 marrow stromal cells initially plated at 500 cells per cm² and incubated for 6 d to obtain high-density P3 cells. The cells were lifted with trypsin/EDTA and assayed on the open stream flow cytometer. **(A)** Uncorrected plot of FS/SS. **(B)** Same sample stained with Annexin V-FITC (R1). **(C)** Same sample after gating out Annexin V⁺ events. A small fraction of very low FS/SS events was Annexin V⁻ debris. Abbreviations: FITC, fluorescein isothiocyanate; FS, forward scatter; SS, side scatter (Figure reproduced with permission from Stem Cells, 2005, 22(5))

8. The cells analyzed by the above method can be sorted using a flow cytometer. Cell preparations are run through open stream FACS instrument (FACSVantage; Becton, Dickinson, Franklin Lakes, NJ). To isolate distinct fractions on the basis of FS and SS (Fig. 6.3), divide the AnnexinV⁻ events into four quadrants on the basis of FS and SS. Then offset the sort gates from the boundaries.

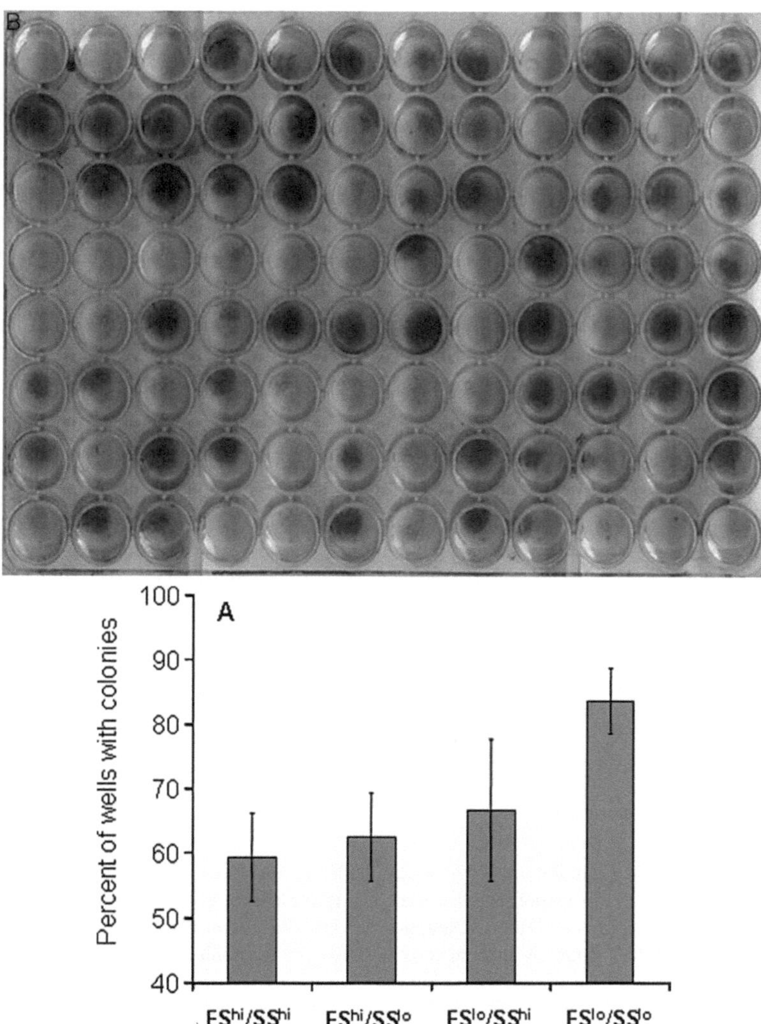

Fig. 6.4 The most clonal cells are characteristically FS^{lo}/SS^{lo}: A: Sc-CFU assay of cells suggesting above 80% colony formation by cells that are low in granularity and size. Crystal Violet Representative microtiter plate in which single FS^{lo}/SS^{lo} cells were deposited. The plate shown in was stained with crystal violet to count colonies after incubation in standard medium for 12–14 d (*See Color Plates*) (Figure reproduced with permission from Stem Cells, 2005, 22(5))

9. The accuracy of sorting single cells into each well of a microtiter plated should be verified routinely by sorting fluorescent beads (Flowchek; Beckman-Coulter) into a test plate and examining the wells with an epifluorescence microscope.
10. The microtiter plates with one MSC per well are incubated in 0.15 mL CCM with change of medium every 4–5 d.
11. After 2 wk, the medium is removed and the wells washed with PBS.
12. Parallel plates could be used for assays of sc-CFUs, for differentiation into osteoblasts, and for differentiation into adipocytes. For sc-CFU assays, the samples were incubated with 1 mL Crystal Violet in methanol for 1 min, washed with water, and colonies with diameters greater than 1 mm counted by microscopy (4× phase contrast) (Fig. 6.4).
13. For assay of differentiation, the microtiter plates were incubated in 0.15 mL per well of αMEM and specific differentiation medium as described in Chapter 7.
14. Cells treated with the Annexin V-FITC were maintained at 4 °C to prevent aggregation and death from the presence of calcium and reagent-induced toxicity.

4 Notes

1. Suspension density for cell counting should be between 0.5 million to 1 million cells per mL.
2. The cells should be well disassociated.
3. The variation in values for log (%G/%T) should be established against samples containing 0.5 or 1 million MSCs per mL when the following parameters were varied: (a) the flow rate was 250, 500 or 900 cells per second; (b) the FS was assayed with 67 or 122 volts and a gain of 2 or with 353 volts and a gain of 1; and (c) the peak for FS of the 20 micron bead was set at 550, 650, or 750; and (d) the peak for SS the 7 micron bead was set at 350, 450 or 550.
4. Confluency of cells in culture is important and cells should be harvested when they are less than 80% confluent.

References

1. Friedenstein, A. J., Chailakhyan, R. K., Latsinik, N. V., Panasyuk, A. F., and Keiliss-Borok, I. V. (1974) Stromal cells responsible for transferring the microenvironment of the hemopoietic tissues. Cloning in vitro and retransplantation in vivo *Transplantation*. **17**, 331–340.
2. Prockop, D. J. (1997) Marrow stromal cells as stem cells for nonhematopoietic tissues *Science*. **276**, 71–74.
3. Colter, D. C., Class, R., DiGirolamo, C. M., and Prockop, D. J. (2000) Rapid expansion of recycling stem cells in cultures of plastic-adherent cells from human bone marrow *PNAS*. **97**, 3213–3218.
4. Smith, J. R., Pochampally, R., Perry, A., Hsu, S. C., and Prockop, D. J. (2004) Isolation of a highly clonogenic and multipotential subfraction of adult stem cells from bone marrow stroma *Stem Cells*. **22,** 823–831.

5. Sekiya, I., Larson, B. L., Smith, J. R., Pochampally, R., Cui, J. G., and Prockop, D. J. (2002) Expansion of human adult stem cells from bone marrow stroma: conditions that maximize the yields of early progenitors and evaluate their quality *Stem Cells*. **20,** 530–541.
6. Prockop, D. J., Sekiya, I., and Colter, D. C. (2001) Isolation and characterization of rapidly self-renewing stem cells from cultures of human marrow stromal cells *Cytotherapy*. **3,** 393–396.
7. Colter, D. C., Sekiya, I., and Prockop, D. J. (2001) Identification of a subpopulation of rapidly self-renewing and multipotential adult stem cells in colonies of human marrow stromal cells. *Proc. Natl. Acad Sci. U S A*. **98,** 7841–7845.
8. Peister, A., Mellad, J. A., Larson, B. L., Hall, B. M., Gibson, L. F., and Prockop, D. J. (2004) Adult stem cells from bone marrow (MSCs) isolated from different strains of inbred mice vary in surface epitopes, rates of proliferation, and differentiation potential. *Blood*. **103,** 1662–1668.
9. Javazon, E. H., Colter, D. C., Schwarz, E. J., and Prockop, D. J. (2001) Rat marrow stromal cells are more sensitive to plating density and expand more rapidly from single-cell-derived colonies than human marrow stromal cells *Stem Cells*. **19,** 219–225.
10. DiGirolamo, C. M., Stokes, D., Colter, D., Phinney, D. G., Class, R., and Prockop, D. J. (1999) Propagation and senescence of human marrow stromal cells in culture: a simple colony-forming assay identifies samples with the greatest potential to propagate and differentiate *Br. J. Haematol*. **107,** 275–281.

Color Plates

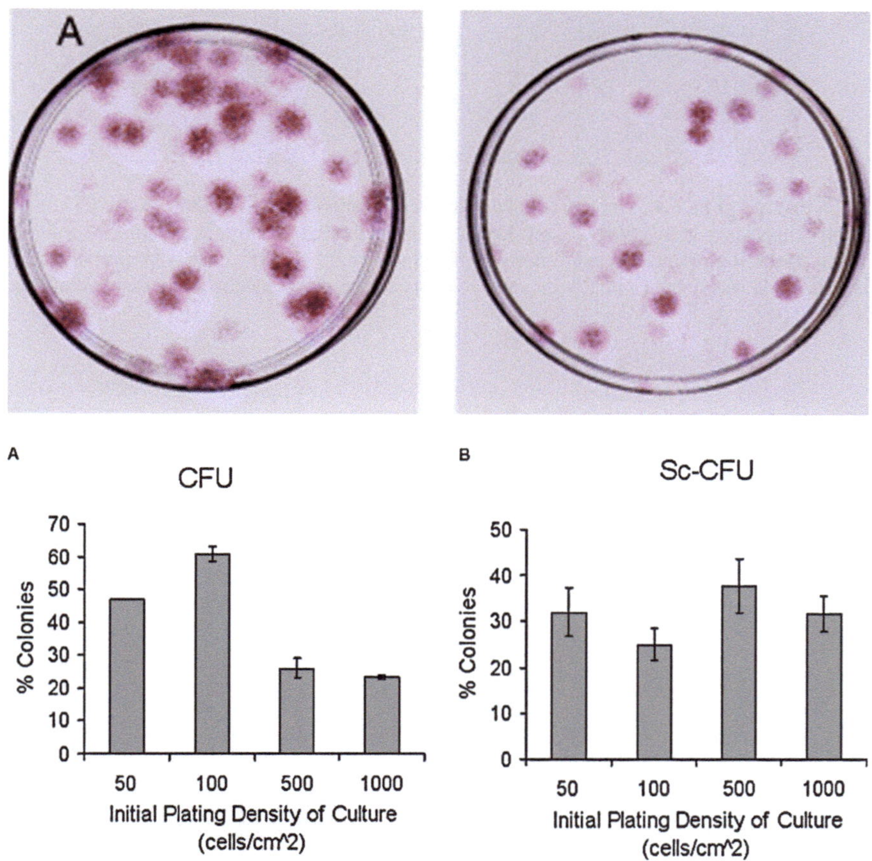

Color Plate 1. **Fig. 6.1** (**A**) Crystal violet stained plates of CFU-F assays performed on two different donors *(10)*. (**B**) Sc-CFU is more sensitive and reproducible than the traditional CFU assay: A: Sc-CFU assay of MSCs initially plated at varying densities and incubated for 10–11 d (mean+/–SD, n=2). B: Standard CFU assay of MSCs plated at same densities and incubated for 13–14 d (mean+/–SD, *n* = 3 or 4)

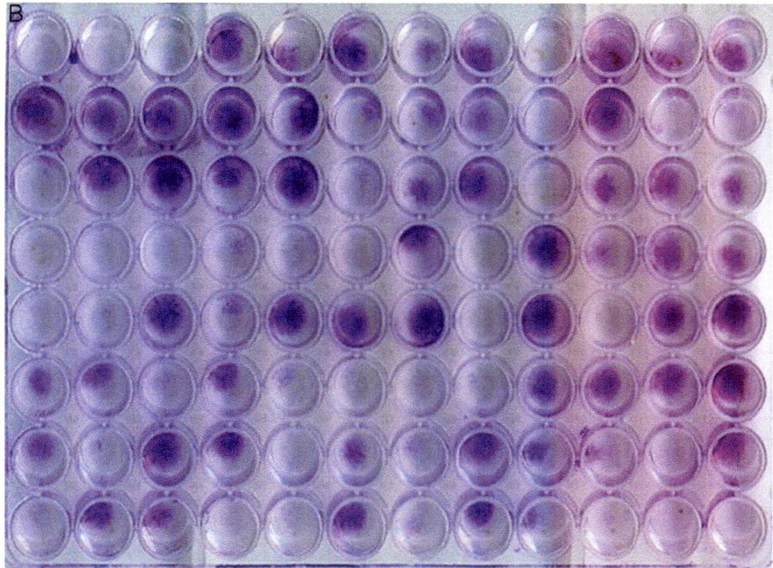

Color Plate 2. **Fig. 6.4** The most clonal cells are characteristically FS^{lo}/SS^{lo}: A: Sc-CFU assay of cells suggesting above 80% colony formation by cells that are low in granularity and size. Crystal Violet Representative microtiter plate in which single FS^{lo}/SS^{lo} cells were deposited. The plate shown in was stained with crystal violet to count colonies after incubation in standard medium for 12–14 d

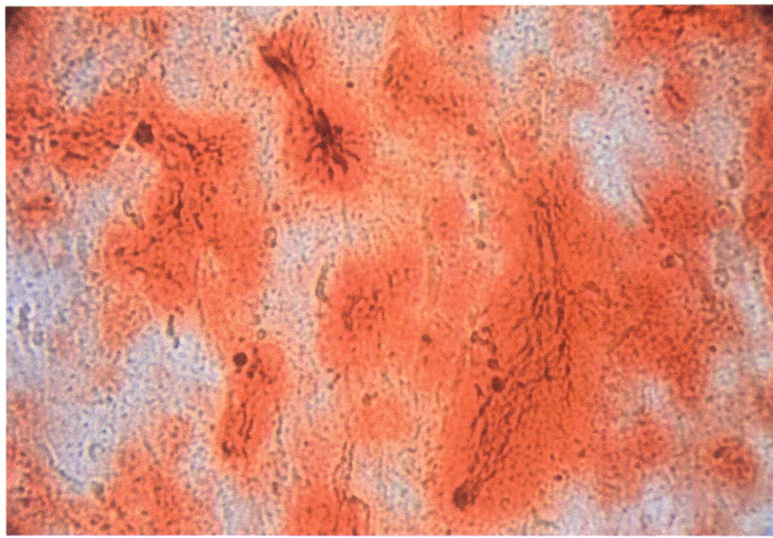

Color Plate 3. **Fig. 7.4** Mineral deposition by MSCs cultured in osteogenic medium indicating early stages of bone formation. Stained with alizarin red S. Mag: 20X

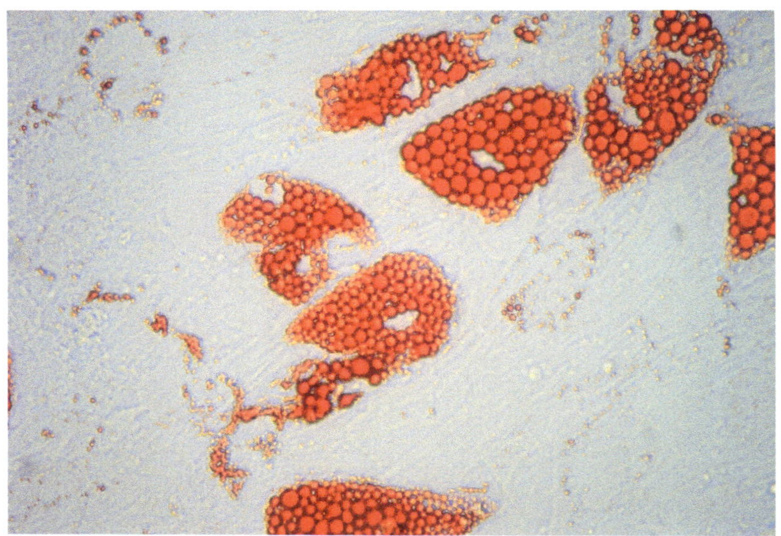

Color Plate 4. **Fig. 7.5** Fat globules seen in MSC culture grown in adipogenic medium indicating differentiating into adipocytes. Stained with oil red O. Mag: 20X

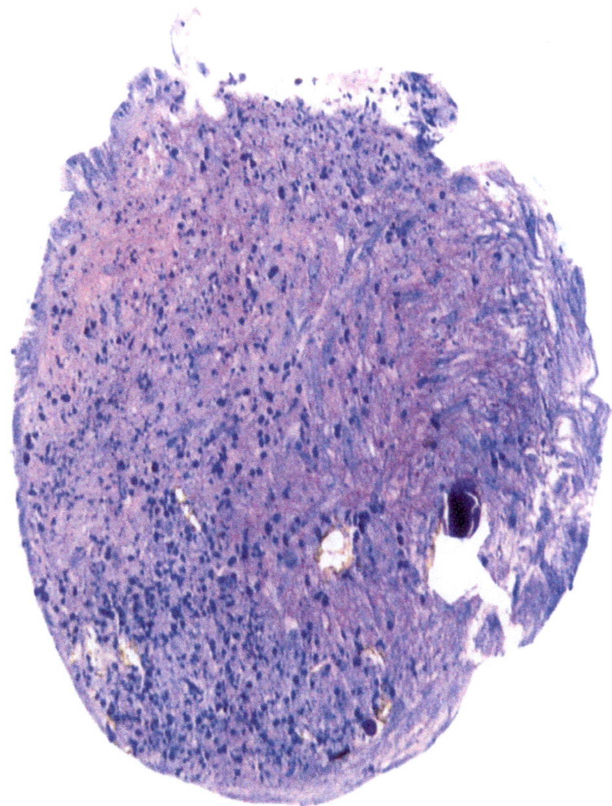

Color Plate 5. **Fig. 7.7** MSC micromass pellet, grown in chondrogenic medium and stained with Toluidine Blue Na Borate. Mag: 10X

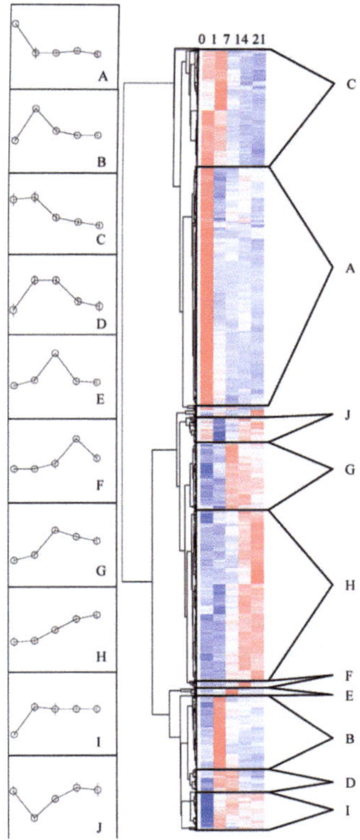

COLOR PLATE 6. **Fig. 10.4** Hierarchial clustering of the genes. Differentially expressed genes (as in Fig. 10.1) for Ch-6 experiment were used to cluster genes hierarchically in dChip program. The red color represents expression level above mean expression of a gene across all samples, the white color represents expression at the mean level, and the blue color represents expression lower than the mean. Co-expressed genes were defined from the clustering picture and average profiles are shown (identified as A-J for Ch-6) on the left of the clustering picture. In the picture a row represents a gene and each column represents a sample from the time course (Day 0, 1, 7, 14, and 21) *(20)*

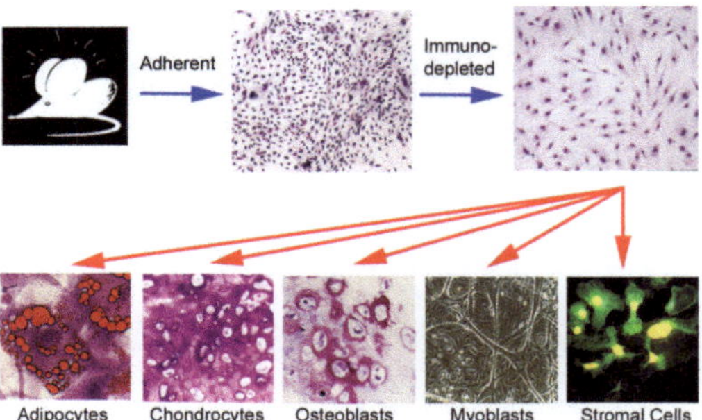

COLOR PLATE 7. **Fig. 12.1** Purification and differentiation of IDmMSCs. The top panel illustrates how immunodepletion removes contaminating hematopoietic lineages from plastic adherent cultures elaborated from murine bone marrow. Images are Geimsa stained plastic adherent populations before and following immunodepletion. The bottom panel shows the potential of IDmMSCs to differentiate into chondrocytes, adipocytes, osteoblasts, myoblasts and hematopoiesis-supporting stroma when cultured under the appropriate conditions in vitro

Chapter 7
Differentiation and Characterization of Human MSCs

Roxanne L. Reger, Alan H. Tucker, and Margaret R. Wolfe

Abstract One of the hallmark characteristics of human MSCs (hMSCs) is their ability to differentiate into adipocytes, chondrocytes and osteocytes in culture. The default fate for hMSCs appears to be bone: if late-passage cultures are left in basic culture medium, the hMSCs will become confluent and produce mineral, an indication of bone formation. However, when grown under certain culture conditions or in media containing specific components, the cells can be driven to become a number of other specific cell types including neural cells, myocytes, and cardiomyocytes. The protocols given here are the basic differentiation procedures for inducing osteogenesis, adipogenesis, and chondrogenesis in cultures of hMSCs. Although there is still no clear consensus on the antigen expression pattern that will define hMSCs, a protocol is also presented for the flow cytometric analysis using a series of antibody panels. The analysis of these surface epitope patterns can aide in the isolation and characterization of hMSCs.

Keywords: MSCs; differentiation; surface epitopes; culture; stem cells; multipotential; characterization.

1 Introduction

The plastic adherent adult stem/progenitor cells from bone marrow originally referred to as fibroblastoid colony forming units, then in the hematological literature as marrow stromal, subsequently as mesenchymal stem cells, and most recently as multipotent mesenchymal stromal cells (MSCs) have generated a great deal of interest in the last few years. Human MSCs (hMSCs) not only are easily obtained from a simple bone marrow aspiration and easily isolated by plastic adherence, but they also have the ability to differentiate into several different cell and tissue types, including bone *(1–3)*, fat *(1,4,5)* and cartilage *(1,6)*. This makes hMSCs an attractive candidate for possible cellular therapies. However, not all MSCs preparations are optimal and not all will differentiate down multiple pathways. To assess various hMSC preparations for potential for multidifferentiation, the cells should be set up

From: *Methods in Molecular Biology, vol. 449, Mesenchymal Stem Cells: Methods and Protocols*
Edited by D.J. Prockop, D.G. Phinney, and B.A. Bunnell

in specific culture conditions to "drive" the MSCs to become specific cell types. Although there are many differentiation procedures depending on the desired type of cell, provided here are the full protocols for simple and basic differentiation into osteocytes, adipocytes, and chondrocytes.

The characterization of hMSCs can be augmented with a battery of monoclonal antibodies and flow cytometry analyses to determine the presence or absence of particular cell surface proteins (Table 7.1). Unlike hematopoietic cells, however, there is little consensus on the antigen expression pattern that can precisely define hMSCs. The use of flow cytometric analysis of hMSCs for surface proteins lies in determining the type of cells obtained and establishing continuity of results among cell preparations and over time in culture. Owing to the large number of antibodies needed to evaluate hMSCs, a very large number of cells are needed to complete the procedure. This problem can be alleviated in part using a panel of antibodies in each analysis thus reducing the number of cells needed from 1 million to 500,000/panel. The number of antibodies that can be mixed is determined by the type of flow cytometer, the number of channels that instrument has available and the cross-reactivity of the antibodies with each other. The list of surface antigens that can be examined on hMSCs is extensive. The protocol presented here includes the antibody panels that we commonly use for basic characterization of hMSC preparations.

2 Materials

2.1 Reagents

1. α minimum essential medium (αMEM) with L-glutamine, without ribonucleosides or deoxyribonucleosides (Invitrogen/GIBCO; catalog # 12561-056).
2. Fetal bovine serum (FBS), premium select, hybridoma qualified, not heat inactivated (Atlanta Biologicals; catalog # S11550) – selected from a screen of 4–5 lots as providing the most rapid growth of hMSCs.
3. L-Glutamine (200 mM) in solution of 0.85% NaCl (Invitrogen/GIBCO; catalog # 25030-081).
4. Penicillin G (10,000 units/mL) and streptomycin sulfate (10,000 μg/mL) in solution of 0.85% NaCl (Invitrogen/GIBCO; catalog # 15140-122) *(Optional).*
5. Phosphate buffered saline (PBS) without Ca^{2+} or Mg^{2+}, 1× (Invitrogen/GIBCO; catalog # 10010-031).
6. 0.25% trypsin and 1 m*M* ethylene diamine tetraacetic acid (EDTA) in Hanks' balanced salt solution (Invitrogen/GIBCO; catalog # 25200-056).
7. 0.4% Trypan blue in solution of 0.85% NaCl (Invitrogen/GIBCO; catalog # 15250-061).
8. Ethanol (70%, 95%, 100%).
9. Isopropanol (100%).
10. Methanol (100%).

Table 7.1 Descriptions of CD markers used for MSC updated flow cytometry panel

MARKER	DESCRIPTION
CD3	OKT3, integral membrane glycoproteins that associates with T cell antigen receptor (TCR), and is required for TCR cell surface expression and signal transduction
CD11b	Aka integrin alpha M, Mac-1; Mediates adhesion to substrates by opsonization with iC3b and subsequent phagocytosis, neutrophil aggregation, chemotaxis
CD14	Aka lipopolysaccharide (LPS) receptor, monocyte differentiation antigen
CD19	Co-receptor with CD21; Earliest B cell antigen in fetal tissue
CD29	fibronectin receptor
CD34	Cell-cell adhesion molecule and cell surface glycoprotein
CD36	Aka platelet GPIV or GPIIIb; thrombospondin receptor. Cell adhesion molecule in platelet adhesion and aggregation, platelet-monocyte and platelet-tumor cell interaction
CD44	Family of cell surface glycoproteins with isoforms generated by alternate splicing of mRNA. Important in epithelial cell adhesion to hyaluronate in basement membranes and maintaining polar orientation of cells; also binds laminin, collagen and fibronectin
CD45	leukocyte common antigen (LCA)
CD49b	very late antigen (VLA) alpha 2 chain - on T cells, Aka GPIa/IIa when expressed on platelets
CD49c	very late antigen (VLA) alpha 3 chain. Receptor for laminin, collagen, fibronectin, thrombospondin
CD49f	Aka very late antigen (VLA) alpha 6 chain; Laminin receptor
CD59	Aka protectin, regulates complement mediated cell lysis by inhibiting formation of membrane attack complex (MAC)
CD73a	Aka ecto-5′-nucleotidase
CD79a	Aka MB-1, B-cell antigen receptor complex associated protein alpha-chain
CD90	Thy-1
CD105	Aka endoglin, regulatory component of TGF-beta receptor complex; mediates cellular response to TGF-beta 1
CD106	Aka VCAM-1; alpha 4 beta 1 ligand
CD117	Aka c-kit, stem cell factor receptor
CD147	Aka neurothelin, extracellular matrix metalloproteinase inducer
CD166	Aka Activated Leukocyte Cell Adhesion Molecule (ALCAM)
CD184	Aka CXCR4, Stromal cell Derived Factor 1 (SDF1). Receptor for the CXC chemokine SDF-1
CD271	Aka Nerve Growth Factor Receptor (NGFR)
HLA-1, ABC	The antigen corresponds to a monomorphic determinant of human HLA class I molecules. HLA-ABC is associated with beta-2 microglobulin.
HLA-2, DR, DQ	All major histocompatibility Class II HLA-DR, DP and most DQ antigens expressed on B cells, antigen presenting cells and activated T cells
Isotype Controls	Mouse IgG1/Mouse IgG2a

11. Deionized (DI) water.
12. Distilled water (dH_2O).
13. Neutral buffered formalin, 10% (NBF) (Sigma; catalog # HT50-1-128).
14. Clear-rite 3 (Richard–Allan Scientific; catalog # 6901).
15. Permount mounting media (Fisher; catalog # SP15-500).

16. CD36 FITC IgG1 (Beckman-Coulter; catalog # IM0766).
17. CD34 PE IgG1 (Beckman-Coulter; catalog # IM1871).
18. CD19 ECD IgG1 (Beckman-Coulter; catalog # IM2708).
19. CD11b PeCy5 IgG1 (Beckman-Coulter; catalog # IM3611).
20. CD45 PeCy7 IgG1 (Beckman-Coulter; catalog # IM3548).
21. CD44 FITC IgG1 (Beckman-Coulter; catalog # IM1219).
22. CD166 PE IgG1 (Beckman-Coulter; catalog # A22361).
23. CD90 PeCy5 IgG1 (Beckman-Coulter; catalog # IM3703).
24. CD49b FITC IgG1 (Beckman-Coulter; catalog # IM1425).
25. CD105 PE IgG1 (Beckman-Coulter; catalog # A07414).
26. CD117 PeCy5 IgG1 (Beckman-Coulter; catalog # IM2657).
27. CD147 FITC IgG1 (BD Biosciences; catalog # 555962).
28. CD49c PE IgG1 (BD Biosciences; catalog # 556025).
29. CD14 ECD IgG2a (Beckman-Coulter; catalog # IM2707).
30. CD29 PeCy5 IgG1 (BD Biosciences; catalog # 559882).
31. CD59 FITC IgG1 (Beckman-Coulter; catalog # IM3457).
32. CD184 PE IgG1 (Beckman-Coulter; catalog # A07409).
33. CD79a PeCy5 IgG1 (Beckman-Coulter; catalog # IM3456).
34. HLA-Class I: ABC FITC IgG1 (BD Biosciences; catalog # 555552).
35. CD271 PE IgG1 (BD Biosciences; catalog # 557196).
36. CD49f PeCy5 IgG1 (BD Biosciences; catalog # 551129).
37. HLA-Class II: DR, DP, DQ FITC IgG1 (BD Biosciences; catalog # 555558).
38. CD73a PE IgG1 (BD Biosciences; catalog # 550257).
39. CD106 PeCy5 IgG1 (BD Biosciences; catalog # 551148).
40. Isotype control IgG1/IgG2a (Beckman-Coulter; catalog # A17599).

2.2 *Equipment*

1. Electric or manual pipet filler/dispenser for mouth-free pipeting of solutions (0.1 to 25 mL).
2. Laminar flow hood (Biosafety Cabinet, Class II).
3. Water bath (37 °C).
4. Tissue culture incubator (37 °C) with controlled and humidified gas with 5% CO_2.
5. Vacuum aspiration source.
6. Hemocytometer with coverslips.
7. General laboratory centrifuge with swinging bucket rotor with buckets and carriers to accommodate various tube sizes (2 mL up to 250 mL).
8. Inverted (phase) microscope for checking culture confluence and counting cells.
9. General purpose laboratory balance (0.01 g sensitivity; 600 g capacity).
10. Multi-channel Flow Cytometer capable of 5 color analysis (Beckman-Coulter FC500 5 Channel Flow Cytometer with HeNe Laser).

11. General laboratory mini-vortex.
12. General laboratory microcentrifuge for 1.5-mL microcentrifuge tubes.

2.3 Supplies

1. Sterile plastic disposable serological pipets: 5 mL, 10 mL, 25 mL, and 50 mL.
2. Sterile plastic disposable pipets for vacuum aspiration.
3. Single channel pipetors, air displacement, capable of accurately measuring from 10 µL–1000 µL, i.e., Eppendorf Research Series 2100 or similar.
4. Sterile aerosol barrier pipet tips, 10, 20, 200, and 1,000 µL.
5. Sterile plastic disposable conical centrifuge tubes: 15 and 50 mL.
6. Plastic disposable snap-cap centrifuge tubes: 1.5 mL.
7. Costar 6-well TC treated microplates (Corning; catalog #3516).
8. Sterile 250-mL filter units 0.22-µm pore size (Millipore, Stericup; catalog # SCGPUO2RE).
9. Sterile 500-mL filter units, 0.22-µm pores (Millipore, Stericup, catalog # SCGPU05PE).
10. Sterile 1,000-mL filter units, 0.22-µm pore (Millipore, Stericup, catalog # SCGPU11PE).
11. Sterile tissue culture dishes/flasks: 15 cm diameter (145 cm^2) dishes (Nunc; catalog # 168381), or T175 (175 cm^2) flasks (Nunc; catalog # 159910).
12. Whatman #1 filter paper.
13. Disposable transfer pipets, 3 mL.
14. 5 mL polystyrene culture test tubes, 12 × 75 mm (Fisher, catalog # 14-961-10) or tubes recommended by the flow cytometer manufacturer.

2.4 Solutions

1. **Complete culture medium (CCM) with 16.5% FBS:** 500 mL αMEM with L-glutamine, 100 mL FBS, 6 mL L-glutamine, 6 mL penicillin/streptomycin (*Optional; see* **Note 1**).

   Filter medium through sterile filter unit. Divide into aliquots you are likely to use for an experiment and store at 4 °C for up to 2 wk. Before an experiment, warm the aliquot to 37 °C.
2. **Osteogenesis differentiation media (ODM):** 192 mL CCM; 10 n*M* dexamethasone (Sigma; catalog # D2915, water soluble—*see* **Note 2**; 200 µL of 1:100 dilution of 1 m*M* stock solution in DI water); 20 m*M* β-glycerol phosphate (Sigma; catalog #G9891; 8 mL of 0.5 *M* stock in CCM); 50 µ*M* L-ascorbic acid 2-phosphate (Sigma; catalog # A8960; 200 µL of 50 m*M* stock solution in DI water).

   Filter medium through sterile filter unit and store at 4 °C for the duration of the osteogeneic culture period.

3. **Adipogenesis differentiation media (ADM):** 200 mL CCM; 0.5 μ*M* dexamethasone (Sigma; catalog # D2915, 100 mg, water soluble – *see* **Note 1**; 100 μL of 1 m*M* stock in DI water); 0.5 μ*M* isobutylmethylxanthine (Sigma; catalog # I5879; 20 μL of 5 m*M* stock in methanol); 50 μ*M* Indomethacin (Sigma; catalog # I7378; 333 μL of 30 m*M* stock in methanol).

   Filter medium through sterile filter unit and store at 4 °C for the duration of the adipogenic culture period.
4. **Chondrogenic media without cytokines (CMwoC)—STOCK:** 500 mL bottle of high-glucose DMEM (GIBCO; catalog # 10569-010) supplemented with: 50 μg/mL L-ascorbic-2-phosphate (Sigma; catalog # A8960; 500 μL of 50 mg/mL stock in DI water); 40 μg/mL L-proline (Sigma; catalog # P0380; 500 μL of 40 mg/mL stock in DI water); 100 μg/mL sodium pyruvate (Sigma; catalog # P5280; 500 μL of 100 mg/mL stock in DI water); 5 mL ITS$^+$ Culture Supplement that consists of 6.25 μg/mL insulin, 6.25 μg/mL transferrin, 6.25 ng/mL selenous acid, 1.25 mg/mL bovine serum albumin, 5.35 mg/mL linoleic acid (B&D Biosciences; catalog # 35-4352).

   CMwoC can be stored for the duration of the chondrogenic culture at 2–8°C.
5. **Chondrogenic media with cytokines (CMwC) – STOCK:** Needed volume of **CMwoC** supplemented with: 10 ng/mL rhTGF-β3 (R&D Systems; catalog # 243-B3; from 10 μg/mL stock in 4 m*M* HCl); 10^{-7}*M* dexamethasone (Sigma; catalog # D2915, 100 mg, water soluble; from 1 m*M* stock in DI water); 500 ng/mL BMP-2 (rhBMP-2, CHO-derived, R&D Systems, catalog # 355-BM OR rhBMP-6, CHO-derived R&D Systems, catalog # 507-BP; from 10 μg/mL stock in PBS, *see* **Note 3**).

   CMwC should be prepared fresh with every use. Stocks of cytokines should be aliquoted and frozen at -20 °C to avoid several freeze-thaw cycles, which can inactivate them.
6. **Alizarin red S stain WORKING (Osteogenesis):** 1 g alizarin red S (Sigma; catalog # A5533); 100 mL DI water.

   Adjust pH of solution between 4.1 and 4.3 using 0.1% ammonium hydroxide. Filter stain through sterile filter unit and store tightly capped at room temperature (RT) protected from light for up to 3 mo.
7. **0.5 Oil red O STOCK:** 2.5 g oil red O (Sigma; catalog # 198196); 500 mL isopropyl alcohol.

   Dissolve completely. Store in a tightly capped bottle at RT protected from light for up to 3 mo.
8. **Oil red-O stain WORKING (Make fresh for each use):** 3 parts 0.5% oil-red-O Stock; 2 parts PBS

   Mix and wait 10 min. Filter stain through sterile filter unit. Wait 10 min before use. Discard any unused stain.
9. **Toluidine blue/1% sodium borate stain - WORKING (Chondrogenesis):** 1 g of toluidine blue (Richard-Allan Scientific; catalog # 90047); 1 g of sodium (Na) borate (Sigma; catalog # S-9640); 100 mL dH_2O.

   First, make 1% Na borate solution (1 g/100 mL dH_2O). Dissolve completely until water is clear. Once clear, add 1 g toluidine blue, dissolve completely. Prefilter using Whatman #1 filter paper and then filter using sterile filter unit. Store tightly capped in an amber bottle for up to 1 mo at RT.

3 Methods

3.1 Plating and Maintaining 6-Well Differentiation Plates for Adipogenesis and Osteogenesis

1. Following the pattern suggested in Fig. 7.1, label 6-well plate with:
   a. Sample number and date
   b. Passage number and cell density per well
   c. 2 wells for bone differentiation medium, 2 wells for fat differentiation medium and 2 wells for control (CCM) medium
   d. Any other pertinent information
2. Add 2 mL of CCM to each well.
3. Add 100,000 cells in a volume less than 2 mL to each well. (*see* **Note 4**)
4. Incubate cells in humidified incubator at 37 °C with 5% CO2.
5. Every 3–4 d before the cells reach 70% confluency, aspirate media from each well, rinse with 2 mL of PBS, and add 2 mL of fresh CCM. Return to incubator.

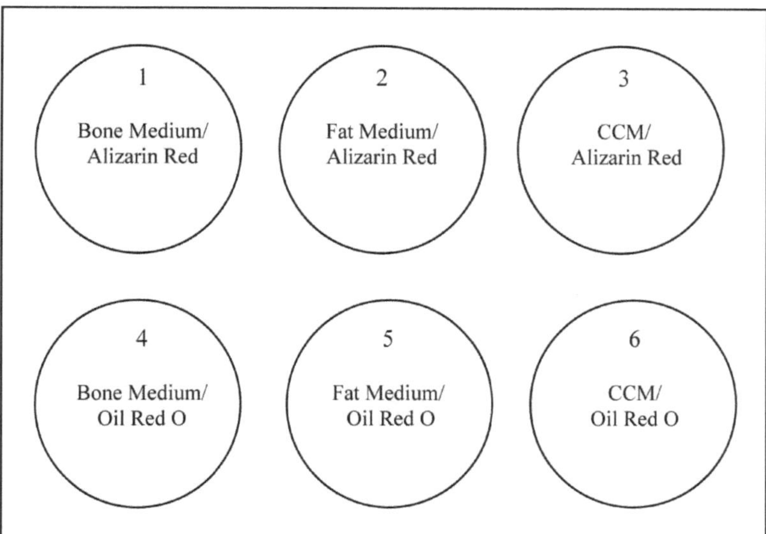

Fig. 7.1 Layout of 6 well plate for differentiation of MSCs. First column of wells (1 and 4) gets MSCs + ODM (bone medium), second column of wells (2 and 5) get MSCs plus ADM (fat medium) and the third column of wells (3 and 6) gets MSCs plus CCM. The first row of wells (1, 2 and 3) gets rinsed with DI water, and then stained with alizarin red S. The second row of wells (4, 5 and 6) gets rinsed with PBS and then stained with oil red O. Alizarin red staining for osteogenesis should be strong in Well 1 and absent to light in wells 2 & 3. Oil red O staining for adipogenesis should be strong in Well 5 and absent to light in the wells 4 & 6. This arrangement allows for controls of media and staining specificity

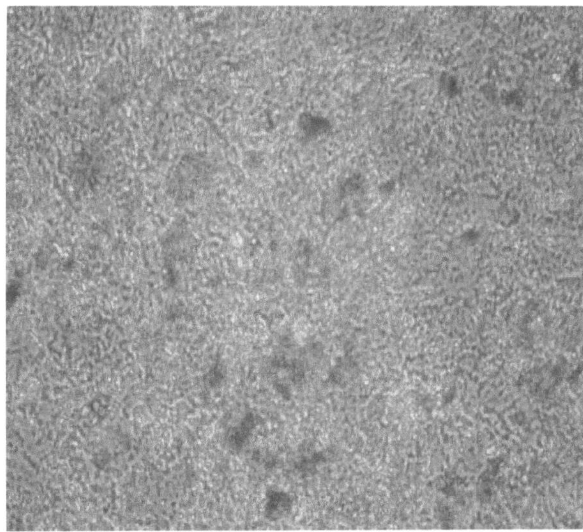

Fig. 7.2 Mineral deposition in MSCs in culture in osteogenic medium as seen under phase contrast. Dark areas indicate mineral that has been manufactured by the cells. Unstained. Mag: 10X

6. After the cells have reached 70% confluency in 2–8 d, aspirate media from each well and rinse with 2 mL of PBS.
7. Add 2 mL of the appropriate media to the wells:
   CCM (Subsection 2.4, 1) to the 2 control wells for no differentiation
   ODM (Subsection 2.4, 2) to the 2 wells for bone/mineral differentiation
   ADM (Subsection 2.4, 3) to the 2 wells for adipogenic differentiation
8. Continue to incubate cells in humidified incubator at 37 °C with 5% CO_2.
9. Every 3–4 d for 21 days wash with 2 mL PBS and replace the appropriate differentiation or control media.
10. Monitor progress of differentiation using the inverted phase microscope. Refer to Figs. 7.2 and 7.3 for appearance of unstained cultures producing fat and mineral.

3.2 *Staining Osteogenic and Adipogenic Differentiation Plates*

1. At the end of 21 d, aspirate media and rinse each well with 2 mL PBS.
2. Add 2 mL of NBF to each well and incubate for 1 h at room temperature.
3. Aspirate NBF from each well and discard.
4. Rinse the wells to be stained with alizarin red S with 2 mL of DI water and aspirate.

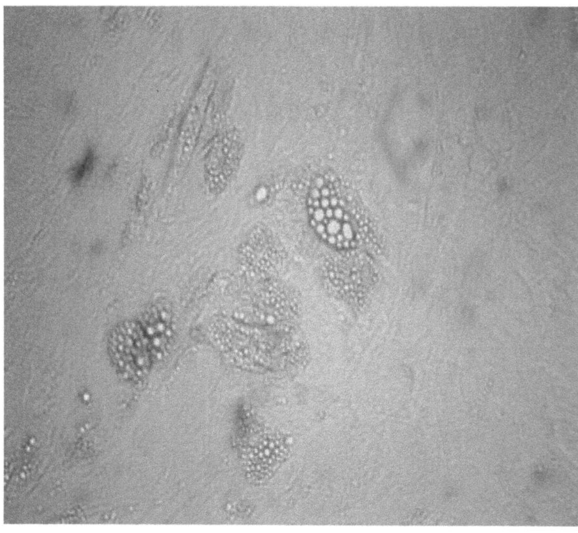

Fig. 7.3 Evidence of fat formation in MSCs cultured in adipogenic medium as seen under phase contrast. Bright round circles are fat globules within the cell. Unstained. Mag: 10X

5. Rinse the wells to be stained with oil red O with 2 mL of PBS and aspirate.
6. Add 2 mL of alizarin red S to each the 3 wells across one row: 1 nondifferentiated well (negative control), 1 fat well (specificity control) and 1 bone differentiated well for actual sample.
7. Add 2 mL of oil red O to each of the 3 wells across one row: 1 nondifferentiated well (negative control), 1 bone differentiated well (specificity control) and 1 fat well for actual sample.
8. Incubate for 20 min at room temperature and then aspirate.
9. Rinse the wells stained with alizarin red S with 2 mL of DI water and aspirate.
10. Repeat 2 times or until background is clear.
11. Rinse the wells stained with oil red O with 2 mL of PBS and aspirate. Repeat two times or until background is clear.
12. Add a final 2 mL of DI water to the alizarin red S stained wells (Wells 1-3).
13. Add a final 2 mL of PBS to the oil red O stained wells (Wells 4-6).
14. Examine plate under inverted microscope for evidence of fat and/or bone differentiation. Negative control wells should not stain at all. Also, fat differentiated cells should not stain with alizarin red S and bone differentiated cells should not stain with oil red O. (*see* **Notes 5** and **6**)

See Figs. 7.2 to 7.5 for illustration of osteogenic and adipogenic cultures, unstained and stained.

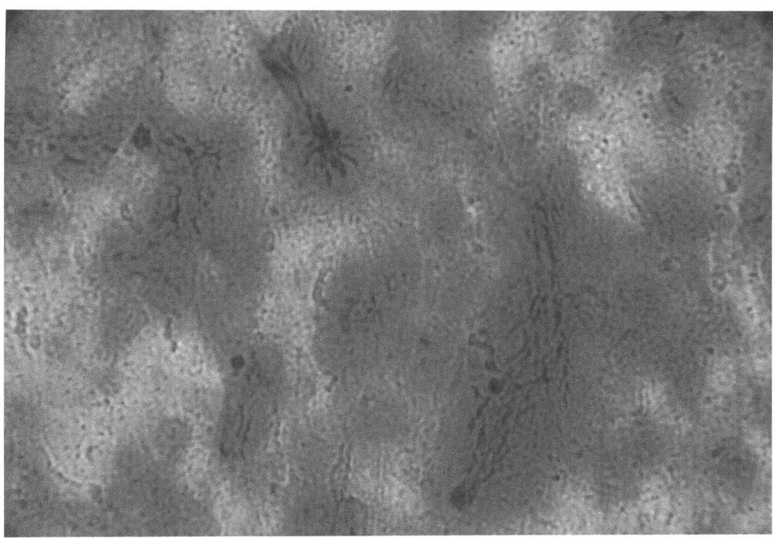

Fig. 7.4 Mineral deposition by MSCs cultured in osteogenic medium indicating early stages of bone formation. Stained with alizarin red S. Mag: 20X (*See Color Plates*)

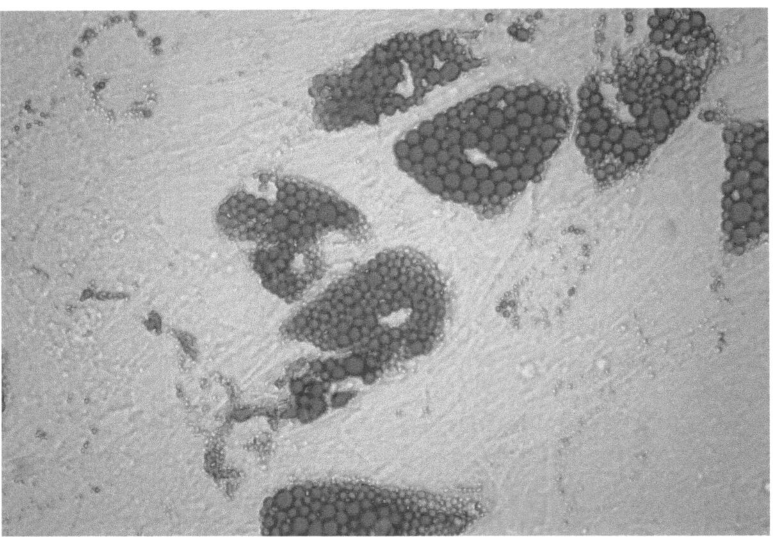

Fig. 7.5 Fat globules seen in MSC culture grown in adipogenic medium indicating differentiating into adipocytes. Stained with oil red O. Mag: 20X (*See Color Plates*)

3.3 Chondrogenesis Differentiation

Harvest cells when 70–80% confluent for this assay. Cells lifted during early to mid-log of growth or those that have reached 100% confluence will not differentiate as well, if at all.

1. Wash harvested MSCs in PBS and resuspend in 1.0 mL CMwC.
2. Do cell count and viability. Adjust to a concentration to 400 viable cells/μL with Chondrogenic Media with Cytokines (CMwC).

For example, if cell count of 1 mL of cell suspension gives 1,000,000 cells/mL, add 1.5 mL CMwC to give 2.5 mL of 400 cells/μL. Thus, 500 μL should contain 200,000 cells.

3. Transfer approx 200,000 MSCs in 500 μL CMwC into a 15-mL conical polypropylene tube.
4. Screw caps on tightly while in hood (sterility is of the utmost importance as no antibiotics are added to the media).
5. Centrifuge the 15-mL conical tube at 450 *g* for 10 min. DO NOT resuspend the pellet and DO NOT aspirate the medium.
6. Place the conicals into the cell culture incubator, which is humidified at 37 °C with 5% CO_2. Loosen the caps on the conicals so that they are simply placed on the conicals without screwing on, allowing for full air exchange. Be sure to screw the caps on tightly before removing the conicals from the incubator. (*see* **Note 7**).
7. Pellets should be visible within 24 h.
8. Change media every 3–4 d by using a P-1000 pipet to remove the old media and add fresh CMwC, paying close attention to detach pellet from plastic with each media change. Take care not to aspirate the pellet when removing old medium.
9. At 21 d, chondrocyte pellet should be 2-4 mm in diameter with BMP-2 or 1–2 mm with BMP-6.
10. At 21 d, chondrocyte pellet may be fixed with NBF, embedded in paraffin, cut into 5-μm sections onto slides, and stained with 1% toluidine blue/1% sodium borate. (*see* **Note 8**).

3.4 Preparing and Staining Paraffin Sections of Chondrogenesis Pellets

1. Deparaffinize in Clear-rite 4 times for 5 min each at RT.
2. Hydrate in descending grades of alcohol from 100%, 95%, dH_2O, 2 × 1 min each at RT.
3. Incubate slides in 1% toluidine blue/1% Na borate solution for 5 min at RT.
4. Rinse in several changes of tap water, until water becomes clear.
5. Rinse slides in dH_2O for 1 min at RT.
6. Dehydrate sections in ascending grades of alcohol from 95% and100%, 2 times for 1 min each at RT.
7. Clear in 4 changes of Clear-rite, 1 min each, at RT.
8. Cover slip in Permount mounting media.
9. Examine by microscopy. See Figures 6 and 7 for illustrations of unstained and stained chondrogenesis pellets.

End Result: purple = cartilage, blue = negative

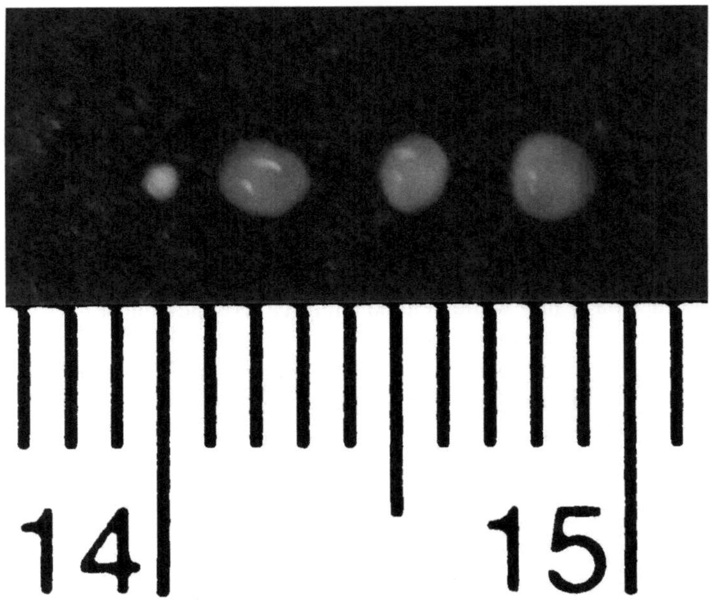

Fig. 7.6 MSC micromass pellets, grown in chondrogenic medium. Scale: mm

3.5 *Staining Procedure for Surface Epitope Characterization of hMSCs*

1. Flow cytometer startup and QC. The instrument procedure for startup and quality control should be performed as described by the manufacturer. This should include the analysis of fluorescent beads to validate the function of the lasers, flow systems, and detection systems. Any problems encountered during this phase should be corrected before proceeding with the analysis of prepared samples.
2. Following the antibody manufacturer's recommendations, the appropriate volume of reagents should be dispensed into a series of eight (8) 1.5-mL microfuge tubes (a panel) as follows:

   Tube 1: CD36 FITC, CD34 PE, CD19 ECD, CD11b PeCy5, CD45 PeCy7
   Tube 2: CD44 FITC, CD166 PE, CD90 PeCy5
   Tube 3: CD49b FITC, CD105 PE, CD117 PeCy5
   Tube 4: CD147 FITC, CD49c PE, CD14 ECD, CD29 PeCy5

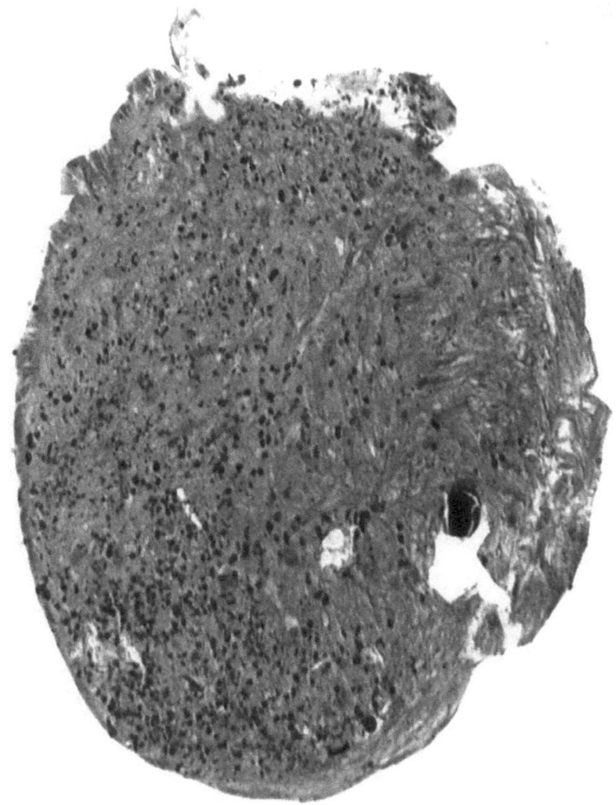

Fig. 7.7 MSC micromass pellet, grown in chondrogenic medium and stained with Toluidine Blue Na Borate. Mag: 10X (*See Color Plates*)

Tube 5: CD59 FITC, CD184 PE, CD79a PeCy5
Tube 6: HLA-Class I: ABC FITC, CD271 PE, CD49f PeCy5
Tube 7: HLA-Class II: DR, DP, DQ FITC, CD73a PE, CD106 PeCy5
Tube 8: Isotype control, IgG1/IgG2a

The tubes containing the antibody cocktails can be made ahead and stored at 4°C in the dark until needed.

3. Harvest cells and count using trypan blue or other method to determine viability. Resuspend cells in PBS at a final concentration of 1×10^6 viable cells/mL. Approx 4×10^6 cells will be required to complete this protocol.
4. Aliquot between 25×10^4 and 5×10^5 cells per tube set-up in Step 3.5, 2. Additionally, set up a ninth tube containing only cell suspension as a control for autofluorescence. Gently vortex to mix and incubate in the dark for 20 min at RT.
5. Wash the cells by adding PBS to the 1.5 mL mark on each tube. Pellet the cells at 100 *g* for 1 min at RT. Remove the supernatant, resuspend the pellet in 1 mL PBS and centrifuge again. Repeat one more time for a total of 3 washes.

Table 7.2 Surface epitope expression on hMSCs at passage 2 for a single donor as determined by flow cytometry[a]

Panel No	Antibody	Expected result on MSCs	hMSCs P2 (% Gated)	Range of % gated for 5 hMSC preps for distribution
1	CD11b - PeCy5	Negative	0.20	ND
	CD19 - ECD	Negative	0.15	ND
	CD34 - PE	Negative	0.45	0.48–1.45
	CD36 - FITC	Negative	0.33	0.27–0.73
	CD45 – PeCy7	Negative	1.54	0.50–1.15
2	CD44 - FITC	Important Pos	74.57	93.68–99.18
	CD90 - PeCy5	Important Pos	99.71	99.32–99.93
	CD166 - PE	Important Pos	99.60	99.30–99.96
3	CD49b - FITC	Positive	25.92	4.84–11.81
	CD105 - PE	Important Pos	98.42	99.49–99.79
	CD117 - PeCy5	Negative	3.13	0.14–0.61
4	CD14 - ECD	Negative	1.52	ND
	CD29 - PeCy5	Positive	96.50	97.61–99.39
	CD49c - PE	Positive	90.89	99.35–99.92
	CD147 - FITC	Positive	97.22	99.27–99.86
5	CD59 - FITC	Positive	99.10	99.81–99.89
	CD79a – PeCy5	Negative	0.85	ND
	CD184 - PE	Positive	2.25	8.46–22.82
6	CD49f - PeCy5	Positive	62.18	56.87–92.78
	CD 271 - PE	Negative	0.70	ND
	HLA-1:ABC - FITC	Positive	96.51	98.93–99.78
7	CD73a – PE	Positive	99.45	ND
	CD106 – PeCy5	Positive	5.84	1.67–18.60
	HLA-class II – FITC	Negative	0.71	ND

[a]The range of % gated of the 5 MSC preparations that are distributed as NIH resource through the NCRR. The expected results are also listed. The unlabeled cells and the isotype controls should be negative. ND = Not Done.

6. Resuspend the final pellet in 500 µL PBS and gently vortex. Be sure no aggregates are present.
7. Using a transfer pipet, place the cell suspensions into the 12 × 75 mm culture test tubes (or recommended device) and analyze on the flow cytometer. Analyze the unlabeled cells (Tube 9) first, followed by the isotype control (Tube 8). Use the results of these 2 control tubes to set the gates and analysis regions. Then read each of the antibody cocktail- labeled cells.

See Table 7.2 for expected expression levels of the panels of antibodies on MSCs and an example of flow cytometry results performed on hMSCs at passage 2 (P2).

4 Notes

1. It is preferable to grow cells without penicillin/streptomycin because any contamination of the culture will not be masked by the presence of antibiotics.

2. Water-soluble Dexamethasone (Sigma D2915) is supplied encapsulated in 2-hydroxypropyl-b-cyclodextrin. Be aware that the actual amount of dexamethasone in the material can vary. Check the label for the actual amount of dexamethasone in your lot and prepare stock accordingly. For example, if the label states that there is 70 mg of dexamethasone per gram of material, to get 3.92 mg of dexamethasone, you would need to weigh out 56 mg of material as supplied.
3. 500 ng/mL BMP-2 may be replaced with 500 ng/mL BMP-6 as cost necessitates. However, BMP-6 does not promote chondrogenesis differentiation as robustly as BMP-2 (data unpublished).
4. If there is an insufficient amount of cells, a lower density can be substituted as long as each well receives the same number of cells.
5. Cultures can be kept in this condition for extended periods to allow for microscopic analysis. Add more DI water or PBS as evaporation decreases the levels.
6. Confluent culture in bone differentiation plates can tend to lift in sheets or form a "ball". The ball can be sliced and made into slides for microscopic analysis for mineral deposition.
7. Before harvesting the cells, place a clean, autoclaved test tube rack that is sufficiently large to hold all of the 15 mL conical tubes and to allow an empty space on all 4 sides of each tube into the incubator. Never remove the rack from the incubator during the course of the assay. Use another clean rack to transport the tubes between the incubator and hood.
8. When staining, it is best to also stain a positive control, known to contain proteoglycans. Chondrogenic pellets successfully differentiated and stained should appear as a positively stained dark purple color, with a negative blue background.

References

1. Baksh, D., Song, L., and Tuan, R. S. (2004). Adult mesenchymal stem cells: characterization, differentiation, and application in cell and gene therapy. *J. Cell. Mol. Med.* **8(3)**, 301–316.
2. Bruder, S. P., Jaiswal, N., and Haynesworth, S. E. (1997). Growth kinetics, self-renewal, and the osteogenic potential of purified human mesenchymal stem cells during extensive subcultivation and following cryopreservation. *J. Cell. Biochem.* **64**, 278–294.
3. Bruder, S. P., Kurth, A. A., Shea, M., Hayes, W. C., Jaiswal, N., and Kadiyala S. (1998). Bone regeneration by implantation of purified, culture-expanded human mesenchymal stem cells. *J Orthop. Res.* **16**, 155–162.
4. Prockop, D. J. (1997). Marrow stromal cells as stem cells for nonhematopoietic tissues. *Science.* **276**, 71–74.
5. Sekiya, I., Larson, B. L., Smith, J. R., Pochampally, R., Cui, J. G., and Prockop, D. J. (2002). Expansion of human adult stem cells from bone marrow stroma: conditions that maximize the yields of early progenitors and evaluate their quality. *Stem Cells.* **20**, 530–541.
6. Sekiya, I., Larson, B.L., Vuoristo, J.T., Reger, R.L., and Prockop, D.J. (2005). Comparison of effect of BMP-2, -4, and -6 on in vitro cartilage formation of human adult stem cells from bone marrow stroma. *Cell Tissue Res.* **320(2)**, 269–276.

Chapter 8
Freezing Harvested hMSCs and Recovery of hMSCs from Frozen Vials for Subsequent Expansion, Analysis, and Experimentation

Roxanne L. Reger and Margaret R. Wolfe

Abstract Human multipotential stromal cells (hMSCs) are easily isolated from bone marrow and can be expanded by up to 200-fold in culture. Cultures of hMSCs are heterogeneous mixtures of stem/progenitor cells and more mature cell types. The proportion of each cell type in a given culture depends on how the cells are maintained. To maintain their stem cell-like qualities, hMSCs should be plated at low seeding densities (60–150 cells/cm^2), lifted when between 60% and 80% confluent and should not be expanded beyond 4–5 passages. Thus, it is useful to establish a frozen bank of early passage cells. hMSCs store well in vapor phase liquid nitrogen (LN_2) and are easily recovered for further expansion. This chapter describes one method of establishing a bank of early passage hMSCs using a seed lot system and the subsequent recovery of hMSCs from frozen stocks. The recovered cells can then be harvested and used for analyses of identification, functionality, in vitro and/or in vivo experimentation, or further expanded.

Keywords MSCs; freezing; recovery; stromal cells; multipotential; culture.

1 Introduction

The plastic adherent adult stem/progenitor cells from bone marrow originally referred to as fibroblastoid colony forming units, then in the hematological literature as marrow stromal, subsequently as mesenchymal stem cells, and most recently as multipotent mesenchymal stromal cells (MSCs) have generated increasing interest in recent years as potential therapeutic agents in a variety of diseases and injuries. Human MSCs (hMSCs) are easily isolated from bone marrow aspirates by their adherence to tissue culture plastic (see Chapter 1, this volume). One characteristic of hMSCs is the ease with which they differentiate into osteoblasts, chondrocytes and adipocytes in culture (see Chapter 7 - Differentiation and Characterization of Human MSCs, this Volume). Others have reported their differentiation into myocytes, cardiomyocytes, skeletal muscle, and other cell types *(1–9)*. However, the proliferation rates and other stem cell-like properties of the hMSCs

From: *Methods in Molecular Biology, vol. 449, Mesenchymal Stem Cells: Methods and Protocols*
Edited by D.J. Prockop, D.G. Phinney, and B.A. Bunnell

gradually change during expansion and, therefore, it is advisable to not expand them beyond 4 or 5 passages. Two morphologically distinct cell phenotypes are seen in early passage, low density cultures: small, spindle-shaped cells that are rapidly self-renewing (RS cells) and large, flat cells that replicate slowly and appear more mature *(10,11)*. The proportion of RS cells remains high for several passages if the cultures are maintained at low density, but the larger cells predominate in later passages *(10–13)*. Cultures enriched for RS cells are obtained if early passage cultures expanded to about 70%, are lifted and replated at low density. If the cultures of hMSCs are allowed to expand to confluence, RS cells are not recovered on replating at low density. The mature hMSCs expand slowly and have less potential for differentiation than RS cells, but retain the ability to differentiate into mineralizing cells and secrete factors that enhance growth of hematopoietic stem cells and perhaps other cell types.

Because of the limited life-span of cultured hMSCs, it is useful to establish a bank of frozen early-passage cells. One way to do this is to follow a seed-lot system (*14*; Fig. 1). This system allows for the establishment of a bank of early passage cells as well as the generation of working stocks of cells. In this system, passage 0 hMSCs are plated at low density into as many plates as possible (or in multilayered Cell Factories; see Chapter 1 this volume). When the cells reach 60–80% confluence, they are lifted and frozen down. A portion of the frozen vials is then set aside as a "seed lot" and are maintained separately so they remain unused and are not handled during retrieval of the working lots. The remaining vials are the "working lot" to be used for on-going research purposes. When all of the working lot is depleted, a vial is retrieved from the seed lot and used to prepare another working lot. The last vial of the seed lot is used to prepare a second seed lot, which will be removed from the initial P0 culture by two passages. By adequately sizing the initial lots, a large number of very early passage hMSCs may be kept for many years. It is important to have a log or database of frozen hMSCs to insure retrieval of the correct sample(s) for recovery and accurate maintenance of the early passage seed lot.

The number of viable hMSCs recovered from a frozen vial is dependent, to a large extent, on how the cells are frozen down. Early passage hMSCs that will be banked should be maintained at low plating densities and lifted for freezing when 60–80% confluent, when the cells are still actively dividing. In addition, the cells should be grown in antibiotic-free medium so as not to mask any contamination. Gentle harvesting procedures including careful control of the amount of time in trypsin also help ensure a healthy stock of cells.

Dimethylsulfoxide (DMSO) at 5% (v/v) provides the best cryoprotection for hMSCs. Timing of the exposure of the hMSCs to DMSO at room temperature is important for successful recovery of the frozen cells. The DMSO needs to penetrate the cells for about 15 min at ambient temperature; but prolonging this contact at ambient temperature will result in reduced recovery at thawing. Therefore, the freezing process should begin about 20–30 min after the cells have been resuspended in the freeze-down medium. A cooling rate of −1 °C works best for hMSCs. This rate of cooling can be achieved with an alcohol-filled container designed to

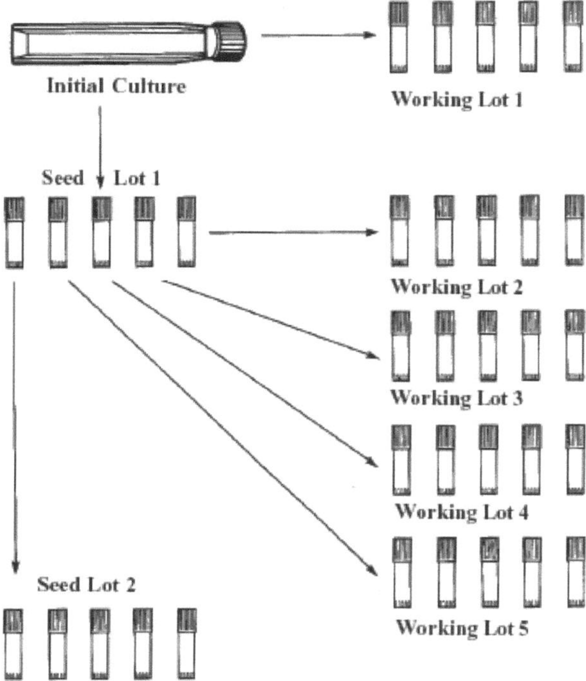

Fig. 8.1 Seed lot system to establish a bank of frozen early passage hMSCs. The seed lot system is one method to ensure that early passage hMSCs are always available for producing new working stock. When preparing the first frozen lot of early passage hMSCs, a portion of the lot is set aside as seed material. The vials designated as seed material are maintained separately from the working stocks to ensure that they remain unused and are not handled during retrieval operations. When the first working stock lot is depleted, a vial is retrieved from the seed lot and used to prepare a second working stock. This continues until all seed vials, except one have been depleted. The last seed vial is then used to prepare a second seed lot. The second seed lot remains only one or two passages from the original material, but may be separated by many years if the lots are adequately sized. (Adapted from *Cryopreservation* Manual: www.nalgenelabware.com/techdata/technical/cryo.pdf).

slowly freeze cells when placed in a low temperature freezer (−80 °C). After freezing overnight at −80 °C, the vials of cells are removed from the container and transferred to LN_2 storage *(1)*.

When removing the vial(s) of MSCs from the LN_2 storage location, carefully check both the label and storage record to ensure that it is the correct vial. Rapid thawing in at 37 °C water bath provides the best recovery; it reduces or prevents the formation of damaging ice crystals within the cells.

The following procedures for freezing down harvested hMSC cultures and for the subsequent recovery of frozen vials of hMSCs have been developed and extensively used in Tulane Center for Gene Therapy lab.

2 Materials

2.1 Reagents

1. α minimum essential medium (αMEM) with L-glutamine, without ribonucleosides or deoxyribonucleosides (Invitrogen/GIBCO; catalog # 12561-056).
2. Fetal bovine serum (FBS), premium select, hybridoma qualified, not heat inactivated (Atlanta Biologicals; catalog # S11550)—selected from a screen of 4–5 lots as providing the most rapid growth of hMSCs.
3. L-Glutamine (200 m*M*) in solution of 0.85% NaCl (Invitrogen/GIBCO; catalog # 25030-081).
4. Penicillin G (10,000 units/mL), and streptomycin sulfate (10,000 μg/mL) in solution of 0.85% NaCl (Invitrogen/GIBCO; catalog # 15140-122) *(Optional).*
5. Phosphate Buffered Saline (PBS), Ca^{2+}- and Mg^{2+}-free, pH 7.40, 1 × (Invitrogen/GIBCO; catalog # 10010-031).
6. 0.25% Trypsin and 1 m*M* ethylene diamine tetraacetic acid (EDTA) in Hanks' balanced salt solution (Invitrogen/GIBCO; catalog # 25200-056).
7. 0.4% Trypan blue in solution of 0.85% NaCl (Invitrogen/GIBCO; catalog # 15250-061)
8. Ethanol (70%).
9. Isopropanol (100%).

2.2 Supplies

1. Sterile plastic disposable serological pipets: 5 mL, 10 mL, 25 mL, and 50 mL.
2. Sterile plastic disposable pipets or Pasteur pipets for vacuum aspiration.
3. Single channel pipetors – must be capable of accurately measuring over a range of 10 μL to 1000 μL.
4. Sterile aerosol barrier pipet tips, 10, 20, 200, and 1000 μL.
5. Sterile plastic disposable conical centrifuge tubes: 15 and 50 mL.
6. Plastic disposable snap-cap microcentrifuge tubes: 1.5 mL.
7. Sterile 250-mL filter units 0.22 μm pore size (Millipore, Stericup, catalog # SCGPUO2RE).*
8. Sterile 500-mL filter units, 0.22 μm pore size (Millipore, Stericup, catalog # SCGPU05PE).*
9. Sterile 1,000-mL filter units, 0.22 μm pore size (Millipore, Stericup, catalog # SCGPU11PE).*
10. Sterile tissue culture dishes/flasks: 15-cm diameter (145 cm^2) dishes (Nunc; catalog # 168381), or T175 (175 cm^2) flasks (Nunc, catalog # 159910).
11. Cryovials: 2.0 mL (Nalgene, Catalog # 5000-0020) or other cryogenic vials for long term storage in LN_2 vapor phase temperatures.
12. Cryo-freezing container (Nalgene, Catalog # 5100-0001) containing 100% Isopropanol.

2.3 Equipment

1. Electric or manual pipet filler/dispenser (0.1–25 mL).
2. Laminar flow hood (biosafety cabinet, Class II).
3. Water bath (37 °C).
4. Tissue culture incubator (37 °C) with controlled and humidified gas with 5% CO_2.
5. Vacuum aspiration source.
6. Hemacytometer with cover slips.
7. General laboratory centrifuge with swinging bucket rotor with buckets and carriers to accommodate various tube sizes (2 mL up to 50 mL).
8. Inverted (phase) microscope for examining cultures and counting cells.
9. −80 °C Freezer (for freezing down vials of harvested hMSCs).

2.4 Stock Solutions

1. **Complete Culture Medium (CCM) with 16.5% FBS:** 500 mL αMEM with L-glutamine, 100 mL FBS, 6 mL L-glutamine, 6 mL penicillin/streptomycin *(Optional)*
   Filter medium through sterile filter unit. Divide into aliquots you are likely to use for an experiment and store at 4 °C for up to 2 wk. Before an experiment, warm the aliquot to 37 °C. (*see* **Note 1**).
2. **Freezing Medium, 1 ×:** 65 mL αMEM with 2 m*M* L-glutamine, 30 mL FBS (same lot as used for cell culture), 5 mL DMSO, 1 mL penicillin/streptomycin (*optional*)
   Filter medium through a sterile filter unit, divide into aliquots and store at −20 °C. (*see* **Note 2**).

3 Methods

3.1 Freezing hMSCs for Long Term Storage

Before freezing, the MSCs should be actively growing to insure maximum health and a good recovery after freezing. Using an inverted microscope, quickly check the general appearance of the culture. It is critical that hMSC cultures are frozen when the cells are 60–80% confluent. Look for signs of microbial contamination. If the MSCs have been grown using antibiotics in the culture medium, it is best if the cultures are maintained antibiotic-free for at least one week prior to freezing to help uncover any culture contaminants that may have been masked by the antibiotics. If desired, the cells can be stored in liquid nitrogen for several years with 30–75% recovery.

1. Pre-label cryovials with cell type, passage #, date, initials, and any other pertinent identifying information. Fill the cryo-freezing container with room temperature isopropanol as directed by manufacturer.
2. If lifting cells from a culture plate or flask, follow steps 3.2.7 to 3.2.14. If harvesting from a Cell Factory, follow steps 3.2.1, 13 through 3.2.1, 28 in Chapter 1, this volume.
3. After counting a sample, concentrate the harvested cells by centrifugation for 10 min at 480 g at room temperature and suspend the pellet in freezing medium to give concentration of about 1 × 106 cells per mL. (*see* **Note 3**).
4. Sterilely aliquot 1 mL of cell suspension into each prelabeled cryovial.
5. Transfer the filled cryovials to the room temperature cryo-freezing container and store overnight to 24 h in a –80 °C freezer.
6. The next day, transfer the cryovials to LN_2 for long-term storage.

3.2 Recovery of Frozen hMSCs

1. In a biosafety cabinet and using aseptic technique, label a 15 cm diameter tissue culture plate (or a T175 flask) for frozen sample recovery with sample number, date, and other information to identify the sample.
2. Add 25 mL of prewarmed (37 °C) CCM to labeled 15-cm diameter plate (30 mL to a T175 flask).
3. Transport the vial from LN_2 in dry ice to a 37 °C water bath. (*see* **Note 4**).
4. Thaw vial of frozen hMSCs in the 37 °C water bath until about 95% thawed. (*see* **Note 5**).
5. Using a 5-mL sterile pipet, add the thawed cells evenly over the surface of the CCM in the 15-cm diameter plate/T175 flask. Gently rock the plate to evenly disperse the cells throughout the media. (*see* **Note 6**).
6. Incubate the cells for 24 h in the tissue culture incubator at 37 °C with humidified 5% CO_2.
7. After 24 h, aspirate the medium, which contains nonadherent/dead cells.
8. To lift the adherent cells, wash plate (or flask) with 25 mL PBS, aspirate and discard. Add 5 mL of prewarmed trypsin / EDTA solution, rock gently to disperse the solution, and incubate for 2–3 min at 37 °C. (*see* **Note 7**).
9. Stop the reaction by adding 10 mL CCM. Gently swirl the media around plate and transfer the detached cells to a sterile 50 mL plastic conical centrifuge tube.
10. Wash any remaining cells with 10 mL PBS and add this wash to the tube containing the detached cells.
11. Centrifuge for 10 min at 480 *g* at room temperature with BRAKE ON.
12. Aspirate the supernatant. Add 1.0 mL CCM to cell pellet and gently but thoroughly resuspend the cells using a 5 mL pipet.
13. Add 20 µL 0.4% Trypan blue solution to a 1.5-mL microcentrifuge tube.
14. Add 20 µL of cell suspension (about 15,000 cells) to the Trypan blue solution, mix thoroughly but gently with a pipet, and count in a hemocytometer.

3.3 *Expansion of Recovered hMSCs*

1. To expand the cells by plating at an initial density of 60–150 cells/cm^2 in a 15-cm diameter dish (145-cm^2 culture area) or T175 flask, dilute 0.1 mL of the cell suspension from the recovery plate (Step 12 in Section 3.2) to a concentration of about 10,000 viable cells per mL.

Example: If hemacytometer count indicates concentration is 500,000 viable cells per mL, dilute 0.2 mL in 9.8 mL of CCM = 10,000 cells/mL.

2. Add 24 mL fresh CCM to the 15-cm dish (30 mL to T175 flask) and then about 8,700 (10,500 for T175 flask) of the diluted viable cells to give 60 cells/cm^2.

Example: If diluted sample contains 10,000 cells per mL, add 0.87 mL of the cell suspension (1.05 mL for T175 flask).

3. Every 3 d, remove media and replace with same volume of fresh CCM.
4. Monitor cells by phase microscopy. When they have reached desired state of expansion (usually in 7–10 d), remove medium, wash cells 1 × with 25 mL PBS, and lift with trypsin/EDTA as in Step 8 of Recovery of Frozen hMSCs (Section 3.2).
5. To expand the cultures further, repeat steps 1 to 4.

4 Notes

1. It is preferable to grow cells without penicillin/streptomycin because any contamination of the culture will not be masked by the presence of antibiotics.
2. An exothermic reaction is produced when DMSO is added to aqueous solutions. Therefore, it is important to add the DMSO to the αMEM first and allow the solution to cool before adding the FBS. Also, it is important to use a sterile filter that is safe for DMSO.
3. Do not allow cells to sit in freezing medium for more than 15 min at room temperature. If you have a large number of cells to process, follow the alternative procedure:

   a. Concentrate cells to about 2×10^6 cells per mL in CCM (2 × cell suspension).
   b. Prepare 2 × freezing medium: 93 mL αMEM with 2 mM L-glutamine, 87 mL FBS, 20 mL DMSO
   c. Pipet 0.5 mL of the 2 × freezing medium into each of the cryovials you will need.
   d. Add to each cryovial 0.5 mL of 2 × cell suspension and then quickly transfer the vials to the freezing container and place in a –80 °C freezer for 24 h. The next day, transfer the vials to LN_2 for long term storage.

4. If vial is recovered from liquid nitrogen, there is a danger of the vial exploding as it thaws. This is because liquid nitrogen may seep into a vial if it is immersed in liquid nitrogen and then the vial explodes as it is thawed because of rapidly expanding LN2 vapor. This usually happens during the first few minutes of thawing. Therefore, use gloves and a face shield while handling thawing vials. Also, unscrew the cap slightly to let any trapped liquid nitrogen escape.
5. Thaw until there is only a small piece of ice remaining in the tube. This keeps the cells cool until the next step. Slow thawing will rupture cells owing to formation of ice crystals.
6. It is not advisable to remove the cryoprotective medium by centrifugation after thawing. The hMSCs are fragile from the trauma of freezing and centrifugation will substantially decrease the viability. It is sufficient to dilute out the DMSO with the CCM in the plate.
7. Trypsinization of the cells must be carefully monitored on an inverted microscope. Begin checking after 1 min at 37 °C and return the cells to the incubator for 1 min intervals as needed.

Stop the incubation in trypsin/EDTA after about 90% of the cells have been lifted. Too long an incubation will kill the cells. The time needed varies with the lot of trypsin/EDTA. Tap sides of dish/flask to help loosen the cells.

References

1. Prockop, D. J., Sekiya, I., and Colter, D. C. (2001). Isolation and characterization of rapidly self-renewing stem cells from cultures of human marrow stromal cells. *Cytotherapy.* **3**, 393–396.
2. Sanchez-Ramos, J., Song, S., Cardozo-Pelaez, F., Hazzi, C., Stedeford, T., Willing, A., Freeman, T. B., Saporta, S., Janssen, W., Patel, N., Cooper, D. R., and Sanberg, P. R. (2000). Adult bone marrow stromal cells differentiate into neural cells in vitro. *Exp. Neurol.* **164**, 247–256.
3. Woodbury, D., Schwarz, E. J., Prockop, D. J., and Black, I. B. (2000). Adult rat and human bone marrow stromal cells differentiate into neurons. *J. Neurosci. Res.* **61**, 364–370.
4. Kotton, D. N., Ma, B. Y., Cardoso, W. V., Sanderson, E. A., Summer, R. S., Williams, M. C., and Fine, A. (2001). Bone marrow-derived cells as progenitors of lung alveolar epithelium. *Development.* ***128***, 5181–5188.
5. Wakitani, S., Saito, T., and Caplan, A. I. (1995). Myogenic cells derived from rat bone marrow mesenchymal stem cells exposed to 5-azacytidine. *Muscle Nerve.* **18**, 1417–1426.
6. Fukuda, K. (2002a). Molecular characterization of regenerated cardiomyocytes derived from adult mesenchymal stem cells. *Congenit. Anom. Kyoto.* **42**, 1–9.
7. Fukuda, K. (2002b). Reprogramming of bone marrow mesenchymal stem cells into cardiomyocytes. *C. R. Biol.* ***325***, 1027–1038.
8. Fukuda, K. (2003). Use of adult marrow mesenchymal stem cells for regeneration of cardiomyocytes. *Bone Marrow Transplant.* **32 Suppl 1**, S25–S27.
9. Barbash, I. M., Chouraqui, P., Baron, J., Feinberg, M. S., Etzion, S., Tessone, A., Miller, L., Guetta, E., Zipori, D., Kedes, L. H., Kloner, R. A., and Leor, J. (2003). Systemic delivery of bone marrow-derived mesenchymal stem cells to the infarcted myocardium: feasibility, cell migration, and body distribution. *Circulation* ***108***, 863–868.
10. Colter, D.C., Sekiya, I., and Prockop, D.J. (2001). Identification of a subpopulation of rapidly self-renewing and multipotential adult stem cells in colonies of human marrow stromal cells. *Proc. Natl. Acad. Sci. U S A.* **98**, 7841–7845.
11. Sekiya, I., Larson, B. L., Smith, J. R., Pochampally, R., Cui, J. G., and Prockop, D. J. (2002). Expansion of human adult stem cells from bone marrow stroma: conditions that maximize the yields of early progenitors and evaluate their quality. *Stem Cells.* **20**, 530–541.
12. DiGirolamo, C. M., Stokes, D., Colter, D., Phinney, D. G., Class, R., and Prockop, D. J. (1999). Propagation and senescence of human marrow stromal cells in culture: a simple colony-forming assay identifies samples with the greatest potential to propagate and differentiate. *Br. J Haematol.* **107**, 275–281.
13. Colter, D. C., Class, R., DiGirolamo, C. M., and Prockop, D. J. (2000). Rapid expansion of recycling stem cells in cultures of plastic-adherent cells from human bone marrow. *Proc. Natl. Acad. Sci. U S A.* **97**: 3213–3218.
14. Cryopreservation Manual, Nalge/Nalgene website: www.nalgenelabware.com/techdata/technical/cryo.pdf.

Chapter 9
Gene Expression Analysis at the Single Cell Level Using the Human Bone Marrow Stromal Cell as a Model: Sample Preparation Methods

Beerelli Seshi

Abstract Recent advances in molecular technology, including gene expression microarray analysis, have allowed researchers to examine global patterns of gene expression at high resolution in populations of cultured cells or tissues. Although these techniques can be applied with great sophistication and are useful for addressing many biological questions in cell populations, it is also of great value to assess gene expression at the level of the single cell. This can be achieved by one of two different approaches: (1) specific cell types can be purified from heterogeneous tissues or cultures using immunological methods such as fluorescence-based or magnetic cell sorting or laser capture microdissection, followed by amplification of target cell nucleic acids, and analysis of expressed genes; or (2) immunohistochemical studies and *in situ* expression studies on identical tissue sections can be used to identify genes or sets of genes whose expression correlates with a morphologically or immunochemically distinct cell-type. Using either approach, the target cell types are identified by their morphological or immunohistochemical properties. This chapter is a primer on using single cell gene expression technology to study human bone marrow stromal cells that express mixed lineage markers. Cytomorphological, cytochemical, and immunocytochemical methods as well as gene expression microarray studies demonstrated that single stromal cells simultaneously express markers associated with osteoblast, fibroblast, muscle, and adipocyte differentiation, suggesting that these stromal cells are mesenchymal progenitor cells that have multilineage differentiation capacity. These data characterize human bone marrow stromal cells as adult stem cells. Because of their pluripotent nature, single cell gene expression technology is particularly critical for characterizing and developing the therapeutic potential of these cells.

Keywords Bone marrow; stromal cell; mesenchymal stem cell; mesenchymal progenitor cell; Wright-Giemsa staining; hematoxylin staining; PAS; Nile Red; Oil Red O; acid phosphatase; alkaline phosphatase; *in situ* hybridization; laser capture microdissection; microarray analysis; single-cell genomics.

From: *Methods in Molecular Biology, vol. 449, Mesenchymal Stem Cells: Methods and Protocols*
Edited by D.J. Prockop, D.G. Phinney, and B.A. Bunnell © Humana Press, Totowa, NJ

1 Introduction

Bone marrow (BM) stromal cells are routinely maintained in culture using one of two well-established culture systems: the Dexter-type system *(1)* or the Friedenstein-type system *(2)*. The media for Dexter-type cultures include hydrocortisone and horse serum, and the cells in a Dexter-type culture are morphologically heterogeneous. Dexter-type cultures support in vitro hematopoiesis, especially myelopoiesis, *(3,4)* and the cells in Dexter-type cultures are considered to be mesenchymal progenitor cells (MPCs) *(5)*. The media for Friedenstein-type cultures do not include hydrocortisone or horse serum and the cells in these cultures have a relatively homogeneous morphology. Friedenstein-type cultures support mesengenesis *(6–8)* and the relatively undifferentiated cells in these cultures, which are frequently called mesenchymal stem cells (MSCs) *(6)*, are less efficient in hematopoietic support than MPCs from Dexter-type cultures *(9)*.

Early studies suggested that Dexter-type cultures promote growth of multiple discrete mesenchymal cell types, including osteoblasts, fibroblasts, muscle cells, and adipocytes. However, our attempts to confirm this result by linking expression of lineage-specific genes to specific cell types were unsuccessful and cytological, cytochemical, and immunocytochemical methods instead revealed that diverse mesenchymal cell markers were coexpressed in individual cells *(5)*. These data provided the first clue that bone marrow stromal cells are mesenchymal progenitor cells with the capacity for multilineage differentiation *(5)*. Subsequently, micro-serial analysis of gene expression (micro-SAGE) technology was used to demonstrate that a single "colony-forming-unit-fibroblast" from a Friedenstein-type culture expressed transcripts normally associated with differentiated bone, cartilage, muscle, epithelium and neural cells *(10)*. However, these data remained controversial because in the aforementioned study, micro-SAGE was only applied to one large colony of approx 10,000 bone marrow fibroblasts, leaving the possibility that the cells were not exclusively clonal in origin, but represented discrete singly differentiated mesenchymal cell types. A later study was carried out using a clonally derived bone marrow stromal cell line that expressed markers from all three germ layers *(11)*, and the results of this study corroborated the earlier conclusion that bone marrow stromal cells have the capacity for multilineage differentiation. In addition, recent studies using fluorescence microscopy show that uncommitted mesenchymal stem cells with no prior exposure to differentiation-inducing stimuli exhibit multilineage gene expression *(12)*. Thus, multilineage gene expression is emerging as the defining molecular hallmark of stem or progenitor cells.

Single-cell gene expression analysis is one approach to demonstrate that bone marrow stromal cells are pluripotent. To this end, single stromal cells were isolated using laser capture microdissection and total RNA was isolated and amplified. Gene expression profiling was conducted on amplified samples from single stromal cells using an Affymetrix oligonucleotide gene chip (U95Av2, 12,625 probe sets; Affymetrix, Inc.). The results demonstrated simultaneous expression in a single cell of diverse lineage-specific RNA transcripts, whose expressions are normally correlated with differentiated cellular phenotypes *(13)*. Interestingly, this suggests that the process of cellular differentiation might require downregulation of many multilineage genes as much as or even more than it requires upregulation of cell lineage-specific genes. In addition, this study demonstrated that Affymetrix microarray analysis at the single cell level is a powerful and viable technology, with important relevance to studies of cellular plasticity and cellular differentiation.

Stromal progenitor cells, like hematopoietic stem or progenitor cells, are rare in bone marrow, occurring at an estimated frequency of 1×10^{-4} nucleated cells *(14)*. Cultured stromal cells are the progeny of a stromal cell, and may differ in properties from the stromal cell itself, for which no definitive assay as yet exists. Pluripotent mesenchymal stem cells have been successfully grown from several tissues/organs, including lung *(15)* and adipose tissue *(16)*. Ultimately, the properties of mesenchymal or other stem and progenitor cells will need to be established by analyzing single cells from fresh tissues, and not from studying cells or populations of cells whose properties may have changed during propagation in culture.

Currently, there are relatively few databases on cell type-specific gene expression, and much of the available data (e.g., GeneCards) *(17)* is based on Northern blot analysis of heterogeneous samples. For example, expression of CD45, the common leukocyte antigen, has been reported in virtually every tissue tested so far (i.e., brain, lung, kidney, and liver), which probably indicates that hematopoietic cells reside in or pass through many of the body's differentiated tissues *(18)*. Thus, the results of gene expression studies on heterogeneous samples must be interpreted with caution. In contrast, immunocytochemistry and *in situ* hybridization in conjunction with traditional histological and/or histopathological assessment of tissue samples are powerful tools that have been used extensively to identify cell type-specific gene expression. Unfortunately, these methods are labor-intensive and only useful for experiments requiring low-throughput. Moreover, although co-expression of high abundance markers can potentially be detected by these methods, it is very difficult to detect coexpression of low abundance markers by these methods. Thus, these labor-intensive methods often produce inconclusive results, and an alternative higher throughput, higher sensitivity approach, such as single cell gene expression analysis, is needed to advance research in this field.

Single-cell genomics *(13,19–22)* promises to be a powerful alternative to the cytology or histology-based methods for analyzing cell type specific-gene expression, as it is a high throughput, sensitive method for simultaneous analysis of expression of thousands of genes in a single cell. As with many cell biological methods, successful application of single cell genomics-based methods depends heavily on the quality of the samples. Thus, this chapter presents sample preparation protocols for single cell analytical procedures in detail, with significant emphasis on cytospin techniques, an important method that has been somewhat underused in previous studies of bone marrow stromal cells.

Fig. 9.1 summarizes cytospin-based protocols and their specific applications, as follows: Wright-Giemsa stain/Diff-Quik stain (morphology), periodic acid-Schiff (PAS) stain (glycogen pools), Nile Red and Oil Red O (lipids), acid phosphatase

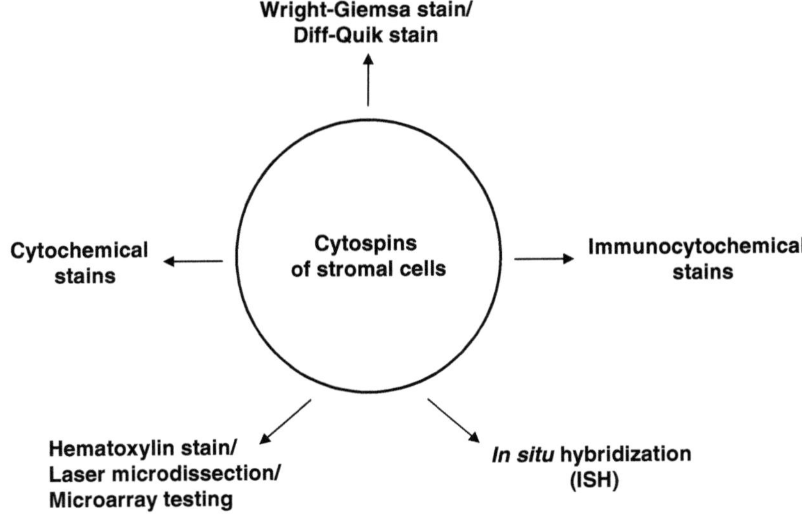

Fig. 9.1 Schematic representation of simple cytospin preparations allowing multidirectional phenotyping of stromal cells

(monocytes), alkaline phosphatase (osteoblasts), and immunoperoxidase detection (hematopoietic and mesenchymal lineage markers) *(5)*. For microarray-based methods, cells are first deposited on a cytospin and stained with hematoxylin. Individual stromal cells are then isolated by laser capture microdissection, and RNA is extracted, amplified, and analyzed on an Affymetrix microchip. Last, note that many methods relevant to single cell genomics can be carried out using high quality commercially available kits. Because such kits are generally self-contained and self-explanatory, they are not discussed here.

2 Materials

2.1 Stromal Cell Culture

1. Basal medium:
    a. McCoy's 5A with L-glutamine and 25 m*M* HEPES buffer (Gibco Invitrogen #12330-031).
2. Medium supplements:
    a. 12.5% fetal bovine serum (Gibco Invitrogen # 26140-079).
    b. 12.5% horse serum (Gibco Invitrogen # 16050-122).
    c. 1 μ*M* hydrocortisone (Sigma # H0135).
    d. 1% penicillin/streptomycin (Gibco Invitrogen: # 15140-122).
3. Phosphate-buffered saline, calcium and magnesium free (PBS-CMF).

4. Cell separation medium (Accu-Prep Lymphocytes) (Accurate Chemical & Scientific Corp.,Westbury, NY).
5. UNI-SEP tubes (UNI-SEP +, 15-mL tube, Cat# U-05; UNI-SEP Maxi +, 50-mL tube, Cat # U-17)) (Accurate Chemical & Scientific Corp.) The UNI-SEP tubes are centrifuge tubes containing a porous insert that allows the passage of erythrocytes, dead cells, and granulocytes, leaving the mononuclear cells above the insert.
6. Table-top centrifuge (Allegra 25R, Beckman Coulter).
7. Tissue culture flasks.
8. Standard plastic centrifuge tubes (50 mL, 15 mL).
9. Marrow aspirate samples (procured per the policies and practices approved for human subjects).

2.2 Percoll Gradient Purification

1. Percoll (Amersham Biosciences, GE Healthcare, Inc).
2. Centrifuge with a fixed angle rotor (Avanti J-25, Beckman Coulter).
3. Clear polycarbonate centrifuge tubes (Beckman # 342080).
4. Standard polystyrene or polypropylene 50-mL centrifuge tubes.
5. Standard polystyrene or polypropylene 15-mL centrifuge tubes.
6. Phosphate-buffered saline, calcium and magnesium free (PBS-CMF).
7. Hanks Balanced Salt Solution, calcium and magnesium free (HBSS-CMF).
8. Na citrate Sigma # S-4641 (Make up 6% stock solution in PBS-CMF).
9. DNase Sigma # D4263-5VL (To reconstitute, add 4 mL PBS-CMF to each vial).
10. Trypsin-EDTA.
11. Fetal bovine serum (FBS).
12. L-Leucine methyl ester hydrochloride (LME) (Aldrich # L-1002 10G).

2.3 Cytocentrifugation

1. Shandon Cytocentrifuge (Thermo Shandon).
2. Disposable cytofunnels: Shandon # 599 1040.
3. Reusable cytospin funnels: Shandon # 599 1021.
4. Cytoclips: Shandon # 599 100 52.
5. White filter cards: Shandon # 599 1022.
6. Cover slips: Fisher #12548B.
7. Slides with no ring; coated: Fisher # 12-550-15 (Superfrost/plus).
8. Slides with ring; no coating: Shandon #599 1051.
9. Slides with no ring; coated: Shandon # 599 1056.
10. Phosphate-buffered saline, calcium and magnesium free (PBS-CMF).

Use slides with coating if planning to apply Wright-Giemsa, Diff-Quik, and cytochemical and immunocytochemical stains. Use slides with no coating if planning to perform microdissection of cells of interest. Slides with ring facilitate easier location of the cells on the slide following cytocentrifugation.

2.4 Wright-Giemsa Stain

1. 100% methanol. Store methanol in flammable cabinets (*see* **Note 1**).
2. Giemsa stain (Sigma #GS1L-1 L). Filter before use. Change daily. Store Giemsa stain in flammable cabinet (*see* **Note 1**).
3. Type I distilled water (deionized water).
4. Wright's buffer: Sigma phosphate buffer, Sigma, cat#: P8165. To prepare the buffer, dissolve the contents in 3.8 L type-I water. Check pH and adjust to 6.4 if necessary and store at room temperature. Good for 1 mo. Alternatively, Wright's buffer can be prepared as outlined in **Note 2**.
5. Working Wright's buffer: Add 5 mL Giemsa stain and adjust volume to 50 mL using Wright's buffer.
6. 80% Ethanol: Using a glass or plastic container, measure 800 mL ethanol and adjust volume to 1,000 mL with Type I water, labeled with the date of preparation and expiration date of 6 mo. Store the container in flame cabinet (*see* **Note 1**).
7. PERMASLIP mounting medium (Item # A325A, Alban Scientific, St. Louis, MO).
8. Xylene (*see* **Note 3**).
9. Slip-Rite Cover Glass No. 1 24 × 50 (Richard-Allan Scientific).
10. Coplin jars.

2.5 Diff-Quik Stain

1. Solution I: Absolute methyl alcohol (VWR or Fisher Scientific).
2. Solution II: Harleco Hemacolor solution 2 (Orange solution) (EM Science, Gibbstown, NJ).
3. Solution III: Diff-Quik solution II (Purple solution) (Dade Behring, Newark, DE).
4. Coplin jars.

See the accompanying labels for information about the dye compositions of the solutions.

2.6 Laser Microdissection

1. Hematoxylin QS (Vector, Burlingame, CA).
2. Diethylpyrocarbonate (DEPC)-treated water.
3. 100% ethanol freshly dispensed from the original container.
4. 50%, 70%, and 90% ethanol freshly prepared using DEPC-treated water.

5. Xylene.
6. Coplin jars.
7. Uncoated glass slides.
8. Shandon cytospin centrifuge.
9. A laser-capture microdissection system (LCM, Arcturus) or Laser Microdissection and Pressure Catapulting system (LMPC, P.A.L.M. Microlaser Systems).

3 Methods

3.1 Stromal Cell Culture

Culture expansion of bone marrow stromal cells includes, a) density gradient isolation of bone marrow mononuclear cells, b) setting up of marrow cultures, and c) harvesting of culture-expanded stromal cells.

3.1.1 Density Gradient Isolation of BM Mononuclear Cells

1. Use UNI-SEP tubes as they allow the marrow sample to be poured directly into the centrifuge tube, avoiding disruption of the cell-separation medium.
2. Add appropriate amounts of cell-separation medium to the UNI-SEP tubes (3 mL if U-05; and 15 mL if U-17).
3. Centrifuge at ~34 *g* for 1 min at 20 °C to transfer the cell-separation medium below the porous insert.
4. Use marrow samples undiluted or preferably diluted 2 parts BM: 3 parts PBS-CMF, final volumes being 4–11 mL if U-05; and 18–25 mL if U-17.
5. Let sample slide along the tube wall. Do not pour it directly onto the insert.
6. Cap and centrifuge at RT and 544 *g* for 24 min with ACC/DEC both set to 0 (i.e., no brake).
7. After centrifugation, first, collect the ring with a transfer pipet. Then, collect the entire remaining contents of the tube above the plastic insert by decanting the solution. Finally, rinse the top of the plastic insert with 5 mL PBS-CMF and add the rinsed solution to the cells collected.
8. Go to the next section, “Setting up of marrow cultures”.

3.1.2 Setting up of Marrow Cultures

1. Adjust the volume of the cell suspension, resulting from the above procedure, to 50 mL using PBS-CMF.
2. Wash by centrifugation at 1,224 *g* for 20 min at 20 °C.
3. At the end of centrifugation, first discard any floating bone marrow fat with the aid of a regular pipet or a transfer pipet. Then, remove the supernatant from the central part of the tube all the way down without disturbing the pellet. Add 5 mL PBS-CMF to mix well.

4. Resuspend the cell pellet in a final volume of 20 mL PBS-CMF. Use 5-mL pipet to resuspend (5×) thoroughly. Take 300 μl for cell counting. Let sample stand on ice until cell counting (either using a Coulter counter or a hemocytometer).
5. Repeat washing by centrifugation. Remove supernatant. Suspend cell pellet in 10 mL or desired volume of complete McCoy's medium.
6. Depending on the cell count, determine the amount of cell suspension to be added to each flask (at a density of $10\text{-}20 \times 10^6$/flask). Add the complete McCoy's medium first and the cell suspension next to the flasks so that the final volume is 20 mL per T-75 flask. (The reason for adding medium first is to avoid small volumes of samples sticking to the plastic).
7. Incubate flasks at 37 °C/ 5% CO_2/ 92% relative humidity (be sure to have the incubator calibrated using an appropriate meter).
8. Demi-feed 2× per week: Take 13 mL out (along with floating cells), and add 13 mL of the fresh medium.

The cells require 2–3 wk to grow to become confluent.

3.1.3 Harvesting of Culture-Expanded Stromal Cells

Unfractionated stromal cells can be harvested by trypsinization of the monolayers. If purification is planned, first kill the macrophages by treating the monolayers with LME, followed by Percoll gradient purification as described next.

3.2 Percoll Gradient Purification

Percoll gradient purification procedure consists of, a) LME treatment of stromal monolayers, aimed at killing the macrophages, b) Trypsinization of LME-treated monolayers for obtaining a single cell suspension, and c) Discontinuous Percoll gradient centrifugation. LME treatment of cultures, followed by Percoll gradient fractionation effectively removes the macrophages that represent about 35% of cells in Dexter cultures.

3.2.1 LME Treatment of Monolayers

If 2 batches of flasks are going to be treated, start the second batch 20 min after the first one.

1. Remove all media from the T-75 flask [or T-150].
2. Add 10 mL of Hank's balanced salt solution (HBSS-CMF) per T-75 [or 20 mL per T-150 flask] and incubate for 3 min.
3. Remove HBSS-CMF.
4. Add 15 mL of McCoy's medium without serum prewarmed to RT. Then add 5 mL of 40-m*M* stock LME solution prewarmed to RT to obtain the desired 10 mM concentration.

5. Incubate at RT for 60 min. At this point the dead macrophages should appear flat and deformed when observed under an inverted phase contrast microscope.

3.2.2 Trypsinization of Monolayers

1. Remove all media from the flasks subjected to LME treatment.
2. Add 10 mL HBSS-CMF /T-75 flask [or 20 mL/T-150 flask].
3. Let stand for 3 min (no more than 10 min) at RT to leach out Ca^{2+} ions.
4. Remove HBSS-CMF from one flask.
5. Add 2 mL/T-75 flask [or 4 mL/T-150 flask] of Trypsin-EDTA; tightly cap the flask; and incubate at 37 °C for 2–3 min.
6. If multiple flasks are being trypsinized, to avoid flasks overlapping allow 2 min between flasks. Handle 2–3 flasks as a batch. Treat with LME 15–20 min apart.
7. Before adding FBS to neutralize trypsin, check under phase contrast microscope to verify most of the cells are completely detached, and appear discrete and show no aggregates (see step 9 on how to break up clumps and obtain single cell suspension).
8. Add 2 mL/T-75 flask [4 mL/T-150 flask] 100% FBS to neutralize trypsin.
9. To consistently obtain single cell suspension, pipet up and down within the flask ×10 using a 5-mL pipet.
10. You may pool cell suspension from several flasks; leave flasks on ice until all flasks are done.
11. Check to see single cell suspension under the scope by dropping 5 µl of cell suspension on a microscope slide.
12. Add DNase and Na Citrate before Percolling to avoid aggregation of cells. After trypsinization of the monolayer as above, add 400 µl of DNase per T-75 flask. Incubate at RT for 30 min with gentle rotation. Then, add 6% Na Citrate (final conc. 0.6%)
13. Load the resulting cell suspension from each T-75 flask on to two Percoll gradient tubes (see next).

3.2.3 Percoll Gradient Centrifugation

1. First, prepare 10%, 20%, 30%, 50%, and 70% solutions of Percoll in PBS-CMF. Then, prepare the gradients by layering 2 mL of each solution, starting with the heaviest solution (70%) at the bottom and ending with the lightest solution (10%) on the top. Gradients will perform well enough, if stored at 4 °C and used within 1 wk.

   Prepare 2 Percoll gradient tubes per T-75 flask [4 tubes per T-150 flask]. Prepare 9 mL of each particular percent solution, if you need 4 Percoll gradient tubes; 36 mL of each percent solution, if you need 16 Percoll gradient tubes. Make the 9-mL solutions using standard 15-mL centrifuge tubes, and make 36-mL solutions using 50-mL centrifuge tubes. Prepare as follows:

2. Mix each tube well and let it stand.

	Using 15mL tubes		Using 50 mL tubes	
70%	6.3 mL undiluted Percoll	2.7 mL PBS-CMF	25.2 mL undilutedPercoll	10.8 PBS-CMF
50%	4.5 mL undiluted Percoll	4.5 mL PBS-CMF	18 mL undiluted Percoll	18 mL PBS-CMF
30%	2.7 mL undiluted Percoll	6.3 mL PBS-CMF	10.8 mL undiluted Percoll	25.2 mL PBS-CMF
20%	1.8 mL undiluted Percoll	7.2 mL PBS-CMF	7.2 mL undiluted Percoll	28.8 mL PBS-CMF
10%	0.9 mL undiluted Percoll	8.1 mL PBS-CMF	3.6 mL Percoll	32.4 mL PBS-CMF

3. Use a 10-mL syringe with 20-G needle or a pipet to layer the gradient solutions in clear polycarbonate tubes 1 $^{1/16}$ × 3 $^{15/16}$ (Beckman # 342080). Start with 70% and then add in decreasing order: 50%, 30%, 20%, and 10% (2 mL each). Check against light to see that gradients are well formed.
4. Add 2 mL of trypsinized single cell suspension (equivalent of ½ of T-75 flask cell suspension) containing 50% FBS, etc. to the top of the gradient.
5. Centrifuge at 800 *g* (RCF) for 15 min at 4 °C in the Beckman Avanti J-25 using the fixed-angle rotor JA-25-15 with the centrifuge precooled to 4 °C and deceleration set to SLOW and acceleration set to MAX.
6. There should be two bands: A light density band at the interface between the serum and the top of the Percoll gradient (which may not be very prominent and is expected to contain mostly macrophages), and a high-density band approximately between the 30% and 50% Percoll layers.
7. When collecting the light density band be sure to remove the serum layer and top part of the 10% Percoll layer.
8. Before recovering the heavy band, remove the serum and the light band. There may also be high density cells that have adhered to the wall of the centrifuge tube, which should be pooled with the high density band.
9. Pool the respective bands from different tubes. Use a 50-mL tube per four Percoll tubes (equivalent of two T-75 flasks). The purpose of using a large tube is to be able to dilute the Percoll during the subsequent washing steps by centrifugation.

3.3 Cytocentrifugation

3.3.1 Cytospin Procedure

1. Cells should be spun onto slides as soon as possible after trypsinization. The cells will begin to break if the time between trypsinization and cytocentrifugation is too long. From past experience, broken cells were observed after 4 h of delay. Keep cells on ice to minimize degradation.
2. Use 10 µL of cell suspension + 90 µl of PBS-CMF; mix gently in advance. Load 100 µL (~5,000 cells) into cytofunnel. If preparing more than one slide, make enough in one tube for all slides needed.

3. Turn on cytocentrifuge and make sure it is balanced.
4. Load the slide and the disposable cytofunnel onto the metal holder. Hold the cytoclip with the bottom down (which has the base for the slide to rest); the top of the clip has a semicircular notch. The slide should be facing up and the writing surface should be at the top. The cytofunnel is placed so that the hole is against the slide. For the reusable cytofunnels, place slide onto the metal holder, followed by a filter card (there is no distinction between top/bottom and front/back). Then, place the reusable cytofunnel with the hole against the slide.
5. Place the slide assembly into the cytospin. Be sure that you have loaded an even number of slides so that the cytospin is balanced.
6. Set at 478 *g* (150 × 10) for 2 min and hit enter.
7. Load 100 μL of PBS-CMF and prerun as under steps 8–10.
8. Load 100 μL of sample.
9. Place lid onto the rotor and lock down.
10. Press start.
11. After centrifugation, remove the slides and air dry (or not dry, as required). Proceed with staining (*see* **Note 8**).

3.4 Wright-Giemsa Stain

The Wright-Giemsa stain is a polychromatic stain consisting of Giemsa stain (thiazine-eosinate solution) in combination with Wright's buffer and is generally used for staining bone marrow smears, peripheral blood smears, etc. It provides excellent discrimination of cell populations within stromal cell cultures as demonstrated on cytospins *(5)*. The morphologic characteristics of the cell populations in Dexter cultures may be summarized as follows: The MPCs are large cells with relatively large irregular nucleus and abundant cytoplasm that is uniquely compartmentalized into ectoplasm and endoplasm. Macrophages are large cells as well; however, they have a very small round bullet-like nucleus and foamy cytoplasm. In contrast, hematopoietic cells are small cells with minimal amount of cytoplasm. The stain can be applied using an automated device or manually. Here we describe a manual method. We have not tested its compatibility with RNA preservation and down-stream microarray testing (for which we used hematoxylin staining, as described under Laser Microdissection).

3.4.1 Staining Procedure

1. Use clean and dry Coplin jars.
2. Add ~50 mL of each reagent to Coplin jars and label each jar appropriately and perform the staining steps as follows:

Container	Reagent	Procedure
Jar #1:	100% methanol	Dip slides for 1 min
Jar #2:	Giemsa stain	Dip slides for 10 min
Jar #3:	Working Wright's buffer	Dip slides for 10 min
Jar #4:	Type I water	Rinse slides ×10
Jar #5:	Type I water	Rinse slides ×10
Jar #6:	80% ethanol	Quickly rinse slides ×1 to remove precipitate & minimize darkness
Jar #7:	Type I water	Rinse slides ×10 to rinse off alcohol

3. Air dry slides, followed by cover slipping.

3.4.2 Cover Slipping of Slides

1. Place labels onto frosted end of slides.
2. Place slides to be cover slipped under hood.
3. Clean slides, if checked with oil, by dipping slides in xylene until oil is removed.
4. Place a drop of PERMASLIP on the slide and place cover glass onto slide.
5. Remove any bubbles and wipe off excess xylene with gauze and let the mounting medium dry.
6. For quality control, see **Notes 4** and **5**.
7. For troubleshooting, see **Notes 6** and **7**.

3.5 Diff-Quik Stain

An alternative to Wright-Giemsa stain is Diff-Quik stain (a registered trademark of American Scientific Products) that uses a mixture of the thiazine dyes methylene blue and azure A in a buffered solution. Diff-Quik protocol is simple and extremely quick as described below and provides the morphologic detail necessary for identifying the cell populations in stromal cell cultures. We have not tested its compatibility with RNA preservation and downstream microarray testing, for which we used hematoxylin staining, described in the next section. The Diff-Quik staining instructions are as follows:

1. Prepare cytospins as described in the cytospin protocol. Air-dry them.
2. Make sure the slides are completely dried (may take about 30 min). If in a rush, use a fan to dry.
3. Add ~50 mL of each reagent to Coplin jars and label each jar appropriately.
4. Dip slide in solution I (Absolute methanol) for 5 s (5 one-second dips). Allow excess to drain.
5. Dip slide in Solution II (Orange solution) for 5 s (5 one-second dips). Allow excess to drain.
6. Dip slide in Solution III (Purple solution) for 5 s (5 one-second dips). Allow excess to drain. If cell buttons are thick, leave slides in Solution III for 10–15 s.

7. Rinse slide with distilled water.
8. Air-dry stained slides, followed by covering with mounting medium and cover slipping, as under Wright-Giemsa staining.

3.6 Laser Microdissection

Laser microdissection of single cells or single colonies of cells is essential for addressing the issue of stromal cell heterogeneity in cultured samples as well as for investigation of rarely occurring stromal progenitors from fresh marrow samples. Two alternative laser microdissection systems are commercially available, i.e., laser capture microdissection (LCM, Arcturus) and Laser Microdissection and Pressure Catapulting (LMPC, P.A.L.M. Microlaser Systems). Both LCM and LMPC use a laser beam for microdissection of the cells. LCM uses a mechanical device to transport the cells into a collection device, whereas LMPC uses photonic force to catapult the selected sample cells against gravity directly into a standard microcentrifuge tube. The first important step before laser microdissection is to prepare cytospins of unpurified or Percoll-purified MPCs by attaching the dispersed BM stromal cells to uncoated glass slides via cytocentrifugation using Shandon cytospin centrifuge, as described above. The next important method is to stain the cells. To ensure preservation of RNA for subsequent amplification and microarray analysis, several special precautions must be observed (*see* **Notes 9–13**).

3.6.1 Hematoxylin Staining

The hematoxylin method described here is a single-step staining method without requiring a special blueing protocol. It provides morphologic detail sufficient for recognition of MPCs, macrophages and lymphocytes, facilitating microdissection of isolated individual cells without compromising the integrity of the RNA. The hematoxylin staining steps are as follows:

1. Arrange staining vessels to facilitate moving through the staining procedure in a timely manner. This should ensure that the cytospin cells do not stand at room temperature for any period until the step using xylene treatment.
2. Fix the cytospins in 95% ethanol for 10 min.
3. Stain for 30 s using Hematoxylin QS.
4. Wash in DEPC-treated water for 10 s.
5. Wash off excess stain from slide by dipping slide for 10 more sec in fresh DEPC-treated water.
6. Dehydrate in increasing concentration of ethanol (50%, 70%, 90% and 100% in Coplin jars) —60 s each.
7. Treat in xylene for 5–10 min.
8. Let xylene completely evaporate for about 20 min in a fume hood.

The cytospins when dry are ready for laser microdissection. If the cells are not going to be dissected immediately, store the slides in a desiccator.

3.6.2 Poststaining Protocols

1. Picture the targeted cells before and after microdissection.
2. Perform laser microdissection of the targeted cells using LCM (Arcturus) or LMPC (P.A.L.M. Microlaser Systems), following the protocols that accompany the respective systems.
3. Extract mRNA from isolated single cells using the PicoPure RNA isolation kit (Arcturus), as described in the accompanying instructions.
4. Amplify RNA using the RiboAmp RNA amplification kit (Arcturus), following the accompanying instructions.
5. Perform Affymetrix microarray analysis using protocols as ongoing in your Microarray Core Laboratory.
6. For additional details and data analysis, see *(13)*.

4 Notes

4.1 Wright-Giemsa Stain

1. Discard methanol, Giemsa stain, Giemsa-Wright buffer in a waste container labeled "Mixed Flammable Waste."
2. An alternative method to prepare Wright's buffer is as follows:
   Weigh 7.68 g Na_2HPO_4 (sodium phosphate dibasic) (reagent grade) and 19.89 g KH_2PO_4 (potassium phosphate monobasic) (reagent grade)
   Adjust volume to 3 L using type-I water. Adjust pH to 6.4 using 1 *N* NaOH or 0.1 *N* HCl. Store the buffer in glass or plastic container at room temperature, labeled date of preparation and date of expiration of 1 mo.
3. Xylene is flammable and a carcinogen. Wear gloves and use under a well vented hood. Store in flame cabinet. Discard old xylene in a waste container.
4. A bone marrow smear or peripheral blood smear may be stained to evaluate the quality of the stain, using the following scale: Grade A: Good stain quality and no precipitate. Grade B: Moderately good stain quality or slight precipitate after re-staining.
5. The methanol should be changed daily. Reagents in stain dish #2–3 should be changed weekly or as needed. The type I water should be changed when it becomes a moderate dark blue. Changing the water more frequently reduces precipitation.
6. If stain quality is light, too blue or pink, check pH of buffer and distilled water. Make up new reagents and change water if necessary.
7. If there is excessive precipitate, redip in fresh 80% ethanol, rinse in fresh distilled water. Dirty slides can also result in increased precipitation.

4.2 Cytocentrifugation

8. The cytospins are generally air-dried, and the cells are fixed and stained according to the cytochemical and immunocytochemical staining protocols. If contemplating laser microdissection and downstream microarray analysis, see the protocols described under Laser Microdissection, which requires special precautions to be observed.

4.3 Laser Microdissection

9. Use DEPC-treated water in all buffers and solutions starting from Hanks' BSS used for washing the monolayers in preparation for trypsinization.
10. Perform cytocentrifugation at low speed (34*g*). Do not use charged slides to prepare cytospins, as it will be difficult to recover cells after microdissection.
11. Cytospins must be stained immediately; therefore, be ready to fix the cytospin slides in 95% ethanol for 10 min and proceed with the staining procedure. Do not allow cells to air dry before fixation!
12. Be sure to use fresh hematoxylin, fresh ethanol and xylene for each staining session.
13. Do not let the sample dry at any point until after the xylene treatment.

Acknowledgments This work was previously supported by a grant from the National Heart, Lung and Blood Institute (NIH grant # R29 HL059683) and is presently supported by a grant from the National Cancer Institute (NIH grant # R21/R33 CA092731).

References

1. Dexter, T. M., Allen, T. D., and Lajtha, L. G. (1977) Conditions controlling the proliferation of haemopoietic stem cells in vitro. *J. Cell Physiol.* **91**, 335–344.
2. Friedenstein, A. J., Gorskaja, J. F., and Kulagina, N. N. (1976) Fibroblast precursors in normal and irradiated mouse hematopoietic organs. *Exp. Hematol.* **4**, 267–274.
3. Liesveld, J. L., Abboud, C. N., Duerst, R. E., Ryan, D. H., Brennan, J. K., and Lichtman, M. A. (1989) Characterization of human marrow stromal cells: Role in progenitor cell binding and granulopoiesis. *Blood.* **73**, 1794–1800.
4. Dorshkind, K., Schouest, L., and Fletcher, W. H. (1985) Morphologic analysis of long-term bone marrow cultures that support B-lymphopoiesis or myelopoiesis. *Cell Tissue Res.* **239**, 375–382.
5. Seshi, B., Kumar, S., and Sellers, D. (2000) Human bone marrow stromal cell: Coexpression of markers specific for multiple mesenchymal cell lineages. *Blood Cells Mol. Dis.* **26**, 234–246.
6. Pittenger, M. F., Mackay, A. M., Beck, S. C., Jaiswal, R. K., Douglas, R., Mosca, J. D., Moorman, M. A., Simonetti, D. W., Craig, S., and Marshak, D. R. (1999) Multilineage potential of adult human mesenchymal stem cells. *Science.* **284**, 143–147.

7. Wakitani, S., Saito, T., and Caplan, A. I. (1995) Myogenic cells derived from rat bone marrow mesenchymal stem cells exposed to 5-azacytidine. *Muscle Nerve.* **18**, 1417–1426.
8. Horwitz, E. M., Prockop, D. J., Fitzpatrick, L. A., Koo, W. W., Gordon, P. L., Neel, M., Sussman, M., Orchard, P., Marx, J. C., Pyeritz, R. E., and Brenner, M. K. (1999) Transplantability and therapeutic effects of bone marrow-derived mesenchymal cells in children with osteogenesis imperfecta [see comments]. *Nat. Med.* **5**, 309–313.
9. Majumdar, M. K., Thiede, M. A., Mosca, J. D., Moorman, M., and Gerson, S. L. (1998) Phenotypic and functional comparison of cultures of marrow-derived mesenchymal stem cells (MSCs) and stromal cells. *J. Cell Physiol.* **176**, 57–66.
10. Tremain, N., Korkko, J., Ibberson, D., Kopen, G. C., DiGirolamo, C., and Phinney, D. G. (2001) Micro-SAGE analysis of 2,353 expressed genes in a single cell-derived colony of undifferentiated human mesenchymal stem cells reveals mrnas of multiple cell lineages. *Stem Cells.* **19**, 408–418.
11. Woodbury, D., Reynolds, K., and Black, I. B. (2002) Adult bone marrow stromal stem cells express germLine, ectodermal, endodermal, and mesodermal genes prior to neurogenesis. *J. Neurosci. Res.* **69**, 908–917.
12. Minguell, J. J., Fierro, F. A., Epunan, M. J., Erices, A. A., and Sierralta, W. D. (2005) Nonstimulated human uncommitted mesenchymal stem cells express cell markers of mesenchymal and neural lineages. *Stem Cells Dev.* **14**, 408–414.
13. Seshi, B., Kumar, S., and King, D. (2003) Multilineage gene expression in human bone marrow stromal cells as evidenced by single-cell microarray analysis. *Blood Cells Mol. Dis.* **31**, 268–285.
14. Galotto, M., Berisso, G., Delfino, L., Podesta, M., Ottaggio, L., Dallorso, S., Dufour, C., Ferrara, G. B., Abbondandolo, A., Dini, G., Bacigalupo, A., Cancedda, R., and Quarto, R. (1999) Stromal damage as consequence of high-dose chemo/radiotherapy in bone marrow transplant recipients. *Exp. Hematol.* **27**, 1460–1466.
15. Noort, W., Kruisselbrink, A., in't Anker, P., Kruger, M., van Bezooijen, R., de Paus, R., Heemskerk, M., Lowik, C., Falkenburg, J., Willemze, R., and Fibbe, W. (2002) Mesenchymal stem cells promote engraftment of human umbilical cord blood-derived CD34 (+) cells in NOD/SCID mice. *Exp. Hematol.* **30**, 870.
16. Zuk, P. A., Zhu, M., Mizuno, H., Huang, J., Futrell, J. W., Katz, A. J., Benhaim, P., Lorenz, H. P., and Hedrick, M. H. (2001) Multilineage cells from human adipose tissue: Implications for cell-based therapies. *Tissue Eng.* **7**, 211–228.
17. GeneCards. (2002) *http://genome-www.stanford.edu/ cgi-bin/genecards/carddisp? PTPRC &search= Y00062&suff=txt*
18. McKinney-Freeman, S. L., Jackson, K. A., Camargo, F. D., Ferrari, G., Mavilio, F., and Goodell, M. A. (2002) Muscle-derived hematopoietic stem cells are hematopoietic in origin. *Proc Natl Acad Sci U S A.* **99**, 1341–1346.
19. Eberwine, J. (2001) Single-cell molecular biology. *Nat. Neurosci.* **4 Suppl**, 1155–1156.
20. Kamme, F., Salunga, R., Yu, J., Tran, D. T., Zhu, J., Luo, L., Bittner, A., Guo, H. Q., Miller, N., Wan, J., and Erlander, M. (2003) Single-cell microarray analysis in hippocampus CA1: Demonstration and validation of cellular heterogeneity. *J. Neurosci.* **23**, 3607–3615.
21. Kawasaki, E. S. (2004) Microarrays and the gene expression profile of a single cell. *Ann N Y Acad. Sci.* **1020**, 92–100.
22. Bengtsson, M., Stahlberg, A., Rorsman, P., and Kubista, M. (2005) Gene expression profiling in single cells from the pancreatic islets of Langerhans reveals lognormal distribution of mrna levels. *Genome Res.* **15**, 1388–1392.

Chapter 10
Assays of MSCs with Microarrays

Joni Ylöstalo, Radhika Pochampally, and Darwin J. Prockop

Abstract The rapid development of microarray technology during the last decade has greatly expanded the ability to define the genes expressed in cells. This chapter will focus on describing the steps required for conducting successful microarray experiments with multipotential stromal cells (MSCs). A complete microarray experiment, using the Affymetrix system, will be described starting from experimental design and ending with examples of data analysis using the dChip program.

Keywords Multipotential stromal cells; MSCs; microarray; gene expression; RNA; real-time PCR.

1 Introduction

Since the first complementary DNA (cDNA) microarray experiments in the mid 90s *(1)*, the microarray field has evolved with great speed. At present researchers can choose from various microarray technologies, like gene expression arrays *(2)*, protein arrays *(3)*, tissue microarrays (TMAs) *(4)*, comparative genomic hybridization arrays (CGHs), and single nucleotide polymorphism (SNPs) analyzing arrays that determine loss of heterozygosity (LOH), or gene copy number changes *(5)*. Probably the most widely used microarrays are gene expression arrays that provide relative measurements of mRNA levels for tens of thousands of genes simultaneously in a biological sample. Since Affymetrix developed the ground breaking short oligonucleotide arrays *(6,7)*, other platforms that use long oligonucleotides or cDNAs have become widely used and companies like Agilent and Amersham have challenged Affymetrix in the field. There have been concerns about the reproducibility of results obtained using different microarray platforms, but recent reports show that with carefully designed and controlled experiments, reproducibility can be ensured to a large degree *(8–11)*.

In the stem cell field, many research groups have used microarray technology to assay the complex transcriptome of various cells. Researchers have tried to identify common genes expressed among stem cells derived from different sources *(12–15)*.

From: *Methods in Molecular Biology, vol. 449, Mesenchymal Stem Cells: Methods and Protocols*
Edited by D.J. Prockop, D.G. Phinney, and B.A. Bunnell

Attempts to use microarrays to define the "stemness" of cells are difficult because a number of complex assumptions are required. However, the assays can be used to identify similarities and differences among different preparations of cells and to follow changes in the patterns of genes expressed as cells are expanded in culture (*see* Chapter 1 and below). Also, microarrays are useful in defining different subpopulations of MSCs *(16,17)* and in following differentiation of MSCs *(18,19)*. One specific application is to identify new candidates for downstream targets for transcription factors known to drive differentiation of MSCs by mining time-dependent changes in the expression profiles during differentiation to several cellular phenotypes *(20)*.

There are several limitations to microarray data. Because the technology depends on nucleic acid hybridization and because the affinity constants for different hybrids vary, the data generated are used primarily to compare differences in the levels of expression of a given gene between two samples. They do not necessarily reflect the relative levels of expression of different genes in the same sample. To make the comparisons between or among samples meaningful, the signal intensities observed for each mRNAs must be normalized. In the Affymetrix oligonucleotide microarrays, multiple matched and mismatched probes for the same mRNA and algorithms for analysis of signal intensities are used to define mRNA as present or absent, but they cannot be fully relied on to rule out the presence of low abundance mRNAs such as some transcription factors. Also, fold-changes between two samples are difficult to compare if an mRNA is called as absent in one sample but present in another. In addition, the potential of cross hybridization of one mRNA with probes for a second are not completely excluded by the use of matched and mismatched probes. Therefore the expression patterns of critical genes must usually be verified by more quantitative assays such as real-time RT-PCR.

In this chapter, the focus will be on designing a microarray experiments using the Affymetrix technology and analysis by the dChip program (Fig 10.1).

2 Materials

2.1 Chemicals

1. Ethidium bromide.
2. Absolute ethanol.
3. 10 × TBE.
4. Agarose.
5. 3 *M* sodium acetate (NaOAc).
6. DEPC-treated water (Ambion).
7. 0.5 *M* EDTA.
8. Pellet Paint (Novagen).
9. Phenol/chloroform/isoamyl alcohol, saturated with 10 m*M* Tris-HCl pH 8.0/1 m*M* EDTA.

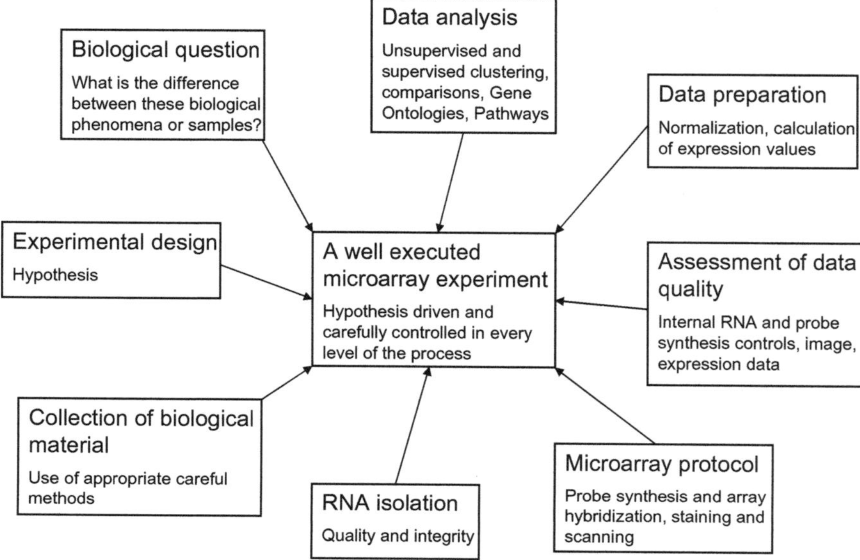

Fig. 10.1 Various aspects of a microarray study

10. Water, molecular biology grade.
11. Water, distilled.
12. Bovine serum albumin (BSA) solution, 50 mg/mL (Invitrogen).
13. Herring sperm DNA (Promega).
14. Control Oligo B2, 3 n*M* (Affymetrix).
15. MES hydrate SigmaUltra (Sigma).
16. MES sodium salt (Sigma).
17. 5 *M* NaCl (RNase-free, DNase-free, Ambion).
18. 0.5 *M* EDTA (Sigma).
19. 10% Tween-20 (Pierce).
20. DMSO (Sigma).
21. R-Phycoerythrin streptavidin (Invitrogen).
22. PBS pH 7.2 (Invitrogen).
23. 20 × SSPE (BioWhittaker).
24. Goat IgG, Reagent Grade (Sigma).
25. Anti-streptavidin antibody (goat), biotinylated (Vector Laboratories).
26. 1 *N* NaOH.
27. 1 *N* HCl.

2.2 Kits

1. RNeasy Mini Kit (Qiagen).
2. GeneChip T7-Oligo(dT) Promoter Primer Kit (Affymetrix).

3. SuperScript Choice System for cDNA Synthesis (Invitrogen).
4. GeneChip In Vitro Transcription (IVT) Labeling Kit (Affymetrix).
5. Sample Cleanup Module (Affymetrix).
6. GeneChip Eukaryotic Hybridization Control Kit (Affymetrix).

2.2 Equipment

1. Agarose gel electrophoresis system with appropriate buffers.
2. UV Spectrophotometer.
3. Bioanalyzer.
4. PCR machine.
5. Microcentrifuge.
6. Hybridization Oven 640 (Affymetrix).
7. Fluidics Station 400 or 450/250 (Affymetrix).
8. GeneChip Scanner 3000 (Affymetrix).

2.4 Supplies

1. Phase Lock Gel (PLG) (Eppendorf Scientific).
2. RNeasy spin columns (Qiagen).
3. Tough-Spots, Label Dots (USA Scientific).
4. Sterile, RNase-free, microcentrifuge tubes, 0.5 and 1.5 mL.
5. Micropipettor.
6. Sterile-barrier, RNase-free pipet tips (pointed, not rounded).

2.5 Reagents

1. 12×MES stock buffer, pH 6.5–6.7, filtered (1.22 *M* MES, 0.89 *M* [Na^+]), 64.61 g of MES hydrate SigmaUltra (Sigma) and 193.3 g of MES Sodium Salt (Sigma) made to 1 L in molecular biology grade water.
2. 2×Hybridization buffer (200 m*M* MES, 2 *M* [Na^+], 40 m*M* EDTA, 0.01% Tween-20), 8.3 mL 12×MES Stock Buffer, 17.7 mL of 5 *M* NaCl (RNase-free, DNase-free, Ambion), 4.0 mL of 0.5 *M* EDTA (Sigma), 0.1 mL of 10% Tween-20 (Pierce) made to 50 mL in water.
3. Wash Buffer A, nonstringent, filtered (6×SSPE, 0.01% Tween-20), 300 mL of 20×SSPE, 1.0 mL of 10% Tween-20 made in to 1 L with water.
4. Wash buffer B, stringent, filtered (100 m*M* MES, 0.2 *M* [Na^+], 0.01% Tween-20), 83.3 mL of 12×MES Stock Buffer, 5.2 mL of 5 *M* NaCl, 1.0 mL of 10% Tween-20 made in to 1 L with water, stored at 2–8 °C and shielded from light.

5. 2×Stain Buffer, filtered (200 m*M* MES, 2 *M* [Na^+], 0.1% Tween-20), 41.7 mL of 12×MES Stock Buffer, 92.5 mL of 5 *M* NaCl, 2.5 mL of 10% made in to 250 mL with water, stored at 2–8 °C and shielded from light.
6. 10 mg/mL Goat IgG Stock (Sigma) in 150 m*M* NaCl, stored at 4 °C.

3 Methods

3.1 Experimental Design

Pilot microarray studies are usually helpful in assessing any unforeseen difficulties. In the pilot experiments it is important to check the variability of the data that will be generated in terms of the variability in the analytical system, in sample preparation, and in the biological systems ***(21)***. The variability in the analytical system of the micorarray chips and detection systems can be checked directly by comparing data from duplicate samples of fully processed and labeled cRNA. The analytical variability in terms of the detection systems and the microarray chip should be small. To minimize it, you must calibrate the equipment and standardize hybridization, washing, staining, and scanning steps using master mixes and various quality controls built into the system. To minimize variability in preparation of samples, it is of great importance to check the quality of the isolated total RNA by either gel electrophoresis and spectrophotometry, or preferably by a bioanalyzer. Also, it is important to assay the final yield of cRNA as a check of the processing steps from synthesis of cDNA through cRNA synthesis and fragmentation. A final check on variability is to compare data from aliquots of the same cell preparation that are processed and assayed in parallel. It is usually adequate to compare duplicate aliquots but triplets may be needed if there are large variations in the data on expression of critical genes.

One critical question is of, course, how much RNA do you need? The standard Affymetrix protocol is based on an assay with one round of linear amplification and the manufacturer of the Affymetrix system recommends 5 µg of total RNA, an amount that can be obtained from about 1 million human MSCs. Starting the RNA isolation with at least 1 million cells or 10 mg of tissue will result typically in at least 10 µg of total RNA, depending on the RNA isolation protocol used. However, our experience indicates that the same quality data can usually be obtained from about 1 µg of good quality RNA. To reduce the amount of RNA required, several manufacturers (Affymetrix, NuGEN, Arcturus, Enzo Lifesciences) have developed protocols to further amplify the RNA using polymerases such as T7 RNA polymerase *(22)* or by PCR of the cDNA. With further amplification, data have been reported with 3 ng of RNA from about 300 cells. However, it is difficult to ensure that all mRNA are amplified equally. Therefore it is frequently important to carry out independent checks on the data.

Biological variability in assays of MSCs arises from differences in cells from different donors and from differences in the MSCs obtained from different bone marrow aspirates taken from different sites in the same donor at the same time, culture conditions such as the medium used, the cell density and cell passage.

3.2 RNA Isolation

1. Harvest MSCs as described in Chapter 1.
2. Start the RNA isolation immediately, or store the cells in RNAlater (Qiagen) at −20 °C or −80 °C for several months. Alternatively, you can also use the cell lysis buffer from the RNeasy Mini Kit and freeze the lysate at −80 °C, or you can snap freeze the samples with liquid nitrogen for storage in −80 °C (*see* **Note 1**).
3. Isolate total RNA from a pellet of human MSCs with a commercially available kit like, RNeasy Mini Kit (Qiagen, Valencia, CA) or RNAqueous Kit (Ambion, Austin, TX). Alternatively, RNA can be isolated from most tissues using TRIzol Reagent (Invitrogen, Carlsbad, CA) followed by clean-up using the RNeasy Mini Kit.
4. For further processing, dissolve the RNA in as small of a volume as possible using DEPC-treated water to ensure high enough concentration for probe synthesis.

3.3 RNA Concentration, Quality and Integrity

1. Determine the RNA concentration and amount by measuring the absorbance at 260 nm (A_{260}) in a spectrophotometer. An absorbance of 1 unit at 260 nm corresponds to 40 µg of RNA per mL measured in water.
2. Determine the quality of the RNA, by measuring the absorbance at 260 and 280 nm, preferably in 10 m*M* Tris-HCl, pH 7.5 to avoid the effect of pH to the absorbance readings. Pure RNA has an A_{260}/A_{280} ratio of 1.9 to 2.1.
3. Check the integrity and size distribution of the purified RNA by agarose gel electrophoresis with ethidium bromide. The 18 S and 28 S ribosomal RNA bands should appear as sharp bands, in a ratio of one-to-two, without any smear, if the RNA has not suffered degradation (*see* **Note 2**).
4. Concentrate the RNA sample if the concentration is less than 0.5 µg/µL by adding 1/10 volume of 3 *M* NaOAc, pH 5.2 and 2.5 volumes of ethanol with 0.5 µL of Pellet Paint (Novagen, San Diego, CA) for easier visualization of the pellet.
5. Mix the contents and incubate at −20 °C for at least 1 h followed by centrifugation at ≥ 12,000 *g* for 20 min at 4 °C.
6. Wash the pellet twice with 80% ethanol and air dry pellet before resuspension to DEPC-treated water.

3.4 Probe Synthesis

The probe synthesis protocol follows the eukaryotic sample alternative protocol for one-cycle cDNA synthesis from 5–20 µg of total RNA (GeneChip Expression

Analysis Technical Manual, Affymetrix, Santa Clara, CA) and uses Superscript Choice System (Invitrogen). Alternatively, One-cycle cDNA Synthesis Kit (Affymetrix) can be used for 1–8 µg of total RNA. In cases where enough RNA is not available, various amplification protocols can be used, like 2-cycle cDNA synthesis (Affymetrix) for 10–100 ng or Ovation Kit (Nugen, San Carlos, CA). Use of Poly-A RNA Control Kit (Affymetrix), so called because of spike-in controls, are recommended to aid in monitoring the entire target labeling process (provide exogenous positive controls). Flow chart of the Affymetrix microarray process is presented in Fig. 10.2.

1. Start the first-strand cDNA synthesis with incubating your purified RNA sample in 10 µL of DEPC-treated water and 2 µL of 50 µ*M* T7-Oligo(dT) Primer (5'-G GCCAGTGAATTGTAATACGACTCACTATAGGGAGGCGG-$(dT)_{24}$-3') in a 200-µL centrifuge tube at 70 °C for 10 min.
2. After incubation, spin the sample, place on ice, and add 4 µL of 5 × First-Strand cDNA buffer, 2 µL of 0.1 *M* DTT and 1 µL 10 m*M* dNTP mix. After mixing, incubate at 42 °C for 2 min.
3. Add 1.0 µL of SuperScript II RT (200 U/µL) for RNA amounts less than 8.1 µg (2.0 µL for 8.1–16.0 µg of RNA and 3.0 µL for 16.1–20.0 µg of RNA). Mix well and incubate at 42 °C for 1 h.
4. Spin the first-strand reaction briefly and place on ice. Add 91 µL of DEPC-treated water, 30 µL of 5 × Second-Strand Reaction Buffer, 3 µL of 10 m*M* dNTP mix, 1 µL 10 U/µL of *Escherichia coli* DNA ligase, 4 µL 10 U/µL of *E. coli* DNA polymerase I and 1 µL of 2 U/µL *E. coli* RNase H. Gently tap the tube to mix before incubating at 16 °C for 2 h.
5. After the incubation, add 2 µL (10 U) of T4 DNA polymerase and incubate at 16 °C for 5 min. Then add 10 µL of 0.5 *M* EDTA (*see* **Note 3**).
6. Purify the double-stranded cDNA by phenol/chloroform extraction (Phase Lock Gel (PLG), Eppendorf Scientific, Hamburg, Germany). Start the purification step by pelleting the PLG by centrifuging at ≥12,000 *g* for 20 s. Then add 162 µL of (25:24:1) phenol:chloroform:isoamyl alcohol (saturated with 10 m*M* Tris-HCl pH 8.0/1 m*M* EDTA) and vortex. Transfer the double-stranded cDNA to the PLG tube. Centrifuge at ≥12,000 *g* for 2 min, and transfer the aqueous upper phase to a fresh tube.
7. Concentrate the sample using ethanol precipitation. If desired, 0.5 µL of Pellet Paint can be added to the solution to make the pellet recovery easier. Add 0.5 volumes of 7.5 *M* NH_4Ac and 2.5 volumes of absolute ethanol (stored at −20 °C) to the sample and vortex.

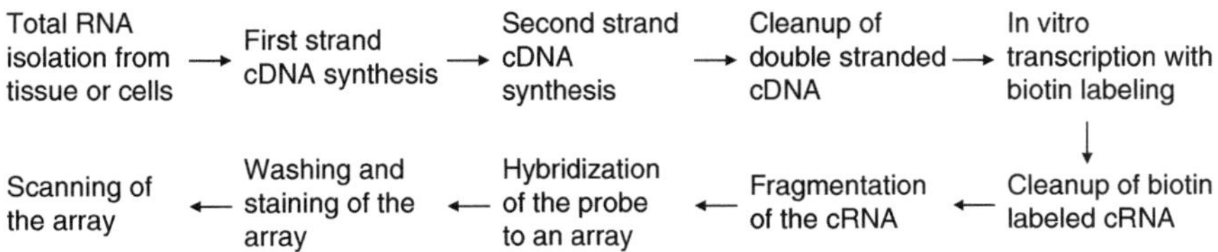

Fig. 10.2 Flowchart for the Affymetrix microarray process

8. Immediately centrifuge the sample at ≥12,000 *g* at room temperature for 20 min.
9. Remove supernatant and wash the pellet twice with 0.5 mL of 80% ethanol (stored at −20 °C) and centrifuge at ≥12000× g at room temperature for 5 min. Air dry the pellet and resuspend it into 12 μL of RNase free water.
10. Use all the double-stranded cDNA generated for the synthesis of biotin-labeled cRNA (GeneChip In Vitro Transcription (IVT) Labeling Kit, Affymetrix) unless the starting amount of total RNA was over 8 μg in which case use half of the generated cDNA and add half water (*see* **Note 4**). Add 8 μL of RNase-free water, 4 μL of 10 × IVT Labeling Buffer, 12 μL of IVT Labeling NTP mix and 4 μL of IVT Labeling Enzyme mix to 12 μL of cDNA in a 200-μL centrifuge tube. Mix the reagents and centrifuge 5 s before 16 h incubation at 37 °C (*see* **Note 5**).
11. Either store the labeled cRNA at −20 °C or −70 °C or proceed to purification immediately.
12. Purify the biotinylated cRNA using RNeasy spin columns (Qiagen), and if needed for fragmentation, concentrate with ethanol precipitation (*see* **Note 6**).
13. Quantify and check the quality of the cRNA by spectrophotometer and calculate the adjusted cRNA yield (adjusted cRNA yield = measured RNA amount – starting amount of total RNA) (*see* **Note 7**).
14. Use 20 μg (adjusted amount) of cRNA for fragmentation to generate 35–200 base cRNA fragments by metal-induced hydrolysis. Add 8 μL of 5 × Fragmentation Buffer to the 20 μg of cRNA (1–32 μL, concentrated with ethanol precipitation if larger volume) (*see* **Note 6**) and add RNase-free water to final volume of 40 μL. Incubate at 94 °C for 35 min before putting the sample on ice.
15. The fragmented target can be checked with gel electrophoresis or the Bioanalyzer and stored at −20 °C or −70 °C for hybridization.

3.5 Hybridization and Scanning

1. For hybridization, add 5 μL of 3 n*M* Control Oligonucleotide B2 (Affymetrix), 15 μL of 20× Eukaryotic Hybridization Controls (Affymetrix), 3 μL of 10 mg/mL Herring Sperm DNA (Promega, Madison, WI), 3 μL of 50 mg/mL BSA (Invitrogen), 150 μL of 2× Hybridization Buffer and 30 μL DMSO (Sigma) to the fragmented cRNA (at least 15 μg) in a 500-μl centrifuge tube. Adjust the final volume to 300 μL with water.
2. Heat the hybridization cocktail to 99 °C for 5 min. During this incubation, wet the room temperature array of choice (Affymetrix) with 1 × hybridization buffer by filling it through one of the septa. Incubate the probe array at 45 °C for 10 min with rotation (Hybridization Oven, Affymetrix).
3. Transfer the hybridization cocktail from 99 °C to 45 °C for 5 min followed by centrifugation for 5 min at ≥12,000 *g*.
4. Remove the buffer from the probe array cartridge and fill with 200 μL of the clarified hybridization cocktail, avoiding any insoluble matter by careful pipetting.

5. Place probe array into the 45 °C Hybridization Oven for 16 h rotating at 60 rpm.
6. Enter the appropriate experiment information to GeneChip Operating Software 1.0 (GCOS, Affymetrix) and set up the fluidics station following the manufacturer's recommendations.
7. After 16 h of hybridization, remove the hybridization cocktail from the probe array and fill it with nonstringent wash buffer (*see* **Note 8**).
8. Prepare the streptavidin phycoerythrin (SAPE) stain solution by mixing 600 μL of 2× stain buffer, 48 μL of 50 mg/mL BSA (Invitrogen), 12 μL of 1 mg/mL SAPE (Invitrogen) and 540 μL of water and divide it to 2.
9. Prepare the antibody solution by mixing 300 μL of 2× stain buffer, 24 μL of 50 mg/mL BSA (Invitrogen), 6 μL of 10 mg/mL Goat IgG Stock (Sigma, St. Louis, MO), 3.6 μL 0.5 mg/mL biotinylated antibody (Vector Laboratories, Burlingame, CA) and 266.4 μL of water.
10. Follow the manufacturer's automated washing and staining fluidics protocol and afterwards make sure no bubbles are present on the array before scanning (*see* **Note 9**).
11. Just before scanning, heat the laser of GeneChip Scanner 3000 (Affymetrix) for 10 min (*see* **Note 10**).
12. Clean the excess fluid from around the septa and apply one Tough-Spot (USA Scientific, Ocala, FL) to each of the two septa if desired. If necessary, clean the glass surface of the probe array with a nonabrasive towel or tissue before scanning the array according to the scanner protocol. The amount of light emitted at 570 nm is proportional to the bound target at each location on the probe array.

3.6 Assessing Data Quality

1. Inspect the probe array image (dat-file) for the presence of image artifacts like scratches, air bubbles or precipitation.
2. The spiked B2 oligo that helps in the identification of the probe area boundaries, serves as a positive hybridization control and is used by the software to place a grid over the image. Check the alignment of the grid by zooming in on each of the four corners and on the center of the image. In some cases you might have to align the grid manually using the B2 oligo binding patterns.
3. The expression report (rpt-file) contains information on different aspects of the microarray quality. The average background and noise (Raw Q) values should be quite similar for all the arrays, but are somewhat equipment dependent. Generally, background values should be less than 100.
4. Poly-A RNA controls, *Bacillus subtilis* genes Dap, lys, phe and thr, can be used to monitor the entire target labeling process as internal controls and should be called "present" on the array with increasing signal values in the order of lys, phe, thr, and dap.

5. The bioB, bioC, and bioD hybridization controls are spiked into the hybridization cocktail and bioB should be called "present" about 50% of the times, whereas bioC, bioD, and cre should always be called "present."
6. Beta-actin and GAPDH can be used as internal controls to assess the RNA sample and assay quality. The signal values of the 3' probe sets are compared to the 5' probe sets and the ratio (3'/5') should not be more than 3. A higher ratio may indicate degraded RNA or inefficient transcription of the double-stranded cDNA or biotinylated cRNA.
7. The percent of probe sets called "present" (30–60%) should be fairly similar between arrays. Extremely low values (less than 20%) are a possible indication of a poor sample quality.
8. The scaling and normalization factors should be comparable among arrays and larger discrepancies may indicate significant assay variability or sample degradation, leading to more noise in the data.

3.7 Data Processing

GCOS defines the probe cells and computes an intensity for each cell after scanning and stores the probe array image as a data file (dat-file).

1. Open the dat-file and the software automatically derives the cell intensity file (cel-file), which contains a single intensity value for each probe cell delineated by the grid (calculated by the cell analysis algorithm).
2. The chip file (chp-file) is generated from the analysis of a probe array and contains qualitative (present, marginal and absent calls) and quantitative (expression values calculated based on perfect matches and mis-matches) analysis for every probe set.
3. Also, a report file (rpt-file) is generated from the chp-file that summarizes the data quality information for a single experiment.
4. Global scaling strategy can be used by setting an average signal intensity of the array to 500, for example. When this is done for all the arrays, they can then be compared using one as a baseline and determine signal log ratio for each gene with the software.
5. Data can then be analyzed further using the Affymetrix software, but we have found it more useful to just use Affymetrix software for generating the image (dat and cel-files) and the absolute calls that will then be used to analyze the data with the dChip program *(23,24)* (*see* **Note 11**).

3.8 Data Preparation for Analysis with the dChip Program

1. Transfer the cel-files (also dat-files can be used) for the arrays to be analyzed, together with the txt-files for each array, containing the absolute calls, to a same

folder for dChip analysis. The image files will not need to be converted if you're using dChip2006, but if the 2005 version is used, use the cel-file conversion tool (Affymetrix) to convert the image from version 4 (binary format) to 3 (text format).
2. Download the library files (CDF-file) for the array used from Affymetrix website, the GeneOntology files from GeneOntology Consortium web-site, and generate the gene information file from these files using the dChip program.
3. Generate "sample information file" that contains the array name, sample name and any other information of your samples like cell passage number or donor identifier.
4. Read in the array data to the dChip program using the files generated. If the array data is on spreadsheet format instead of images, the program can read in the txt-files as external data files for further analysis. The program generates DCP files from the images and txt-files for each array for easier access in future. In cases were the data was generated from different generation arrays, you can use the common probe set files to analyze only the genes that are shared between the arrays.

3.9 Data Analysis with the dChip Program

Only certain aspects of the dChip program are discussed here. For more thorough explanations of the capabilities of the dChip, refer to the dChip manual.

Normalization is needed to adjust the brightness of the arrays to a comparable level. Generally in our lab we use the array with median overall intensity as the baseline array against which other arrays can be normalized at the probe intensity level. Make sure the array you are using as a baseline does not have any problems and is of good quality. The normalization plots can be checked with dChip and with the addition of R. Images can be viewed with dChip together with PM/MM data, and if a baseline array has image contamination, normalization can be performed again using another array. Also, local contaminations can be corrected using the tool in the dChip program. Next step is to calculate model-based expression values, which can be done using either the PM/MM-model or PM-only model that uses only the PM probes. We use the PM/MM-model and we generally truncate the low-values (negative) to 1, but also other options can be used for low-values like truncating them to 10th percentile of expressions called absent or to 0. In addition, the expression values can be log transformed if desired. After the expression value calculation, also array and single outlier analysis results can be viewed. One array is cross-referenced with other arrays through modeling approach to identify problematic arrays. If the arrays have more than 5% outliers, the quality has to be checked because higher outlier percentages are owing to image or sample contaminations. Sometimes the biological difference of the samples can be so large that array outlier numbers are bigger, but if the outlier percentage is higher than 15, discard the array from analysis. After the calculation of the expression values, the data are ready for high-end analysis and the expression values can be exported for viewing on Excel (Microsoft).

Use the array list file to specify which samples are used in each analysis process. In this file, also replicates can be specified.

Microarray studies are useful in finding the genes that are differentially expressed between two samples or sample groups. This can be done by using compare samples function of the dChip program. Select the samples, or groups of samples you want to compare, and set the filtering criteria. We have found it useful to use filtering criteria of at least twofold change and 90% confidence bound of fold change. This results in genes that are reliably different between the selected samples. In addition, genes that have extremely low values, can be filtered out for more accurate interpretation.

To identify specific gene expression patterns across multiple time points and/or patterns between samples, one can perform unsupervised clustering. To obtain genes that show large variation across samples, it is desirable to filter genes so that the results are not affected by the noise from absent or nonchanging genes. In dChip, one can filter genes by their coefficient of variation (CV = ratio of the standard deviation and the mean of a gene's expression value) and by present call percentage. We generally use these criteria to obtain about 3,000 genes that change the most for clustering. Often this will result from CV values higher than 0.3 and present call percentage higher than 25, but these parameters are experiment dependent and can be tuned by trial and error. To obtain genes for supervised clustering, one can filter genes with analysis of variance (ANOVA) by selecting a factor present in the sample information file and filtering genes that are different by certain p-value, for example at least 0.01. Also, correlation filtering can be performed by dChip to obtain genes that have a specific pattern (identified in clustering view).

After a set of genes that are different between samples is obtain, the genes or samples can be clustered. dChip uses hierarchical clustering algorithm *(25,26)* that standardizes the expression values for each gene by linearly adjusting their values across all samples to a mean of zero with a standard deviation of one. Individual genes can then be clustered using an algorithm in dChip program that determines the correlation coefficients (r values) for the normalized expression values (distances between genes were defined as 1 – r). Genes with the shortest distances between them are merged into super-genes, connected in a dendogram by branches with lengths proportional to their genetic distances (centroid-linkage). This process is repeated n-1 times until all genes have been clustered. A similar algorithm is also used to cluster the samples. Instead of hierarchical clustering, some programs might use correlation clustering, K-means or other clustering algorithm. We have found it useful to use hierarchical clustering to cluster the samples with unfiltered set of genes and then trying various filters to get the best separation between different samples. An example of hierarchical clustering of the samples is shown in Fig. 10.3 (reproduced with permission). These sample clusters aid in determining how well the replicates in various levels performed and also test the sets of signature genes that explain the difference between two or more classes of samples. In the clustering

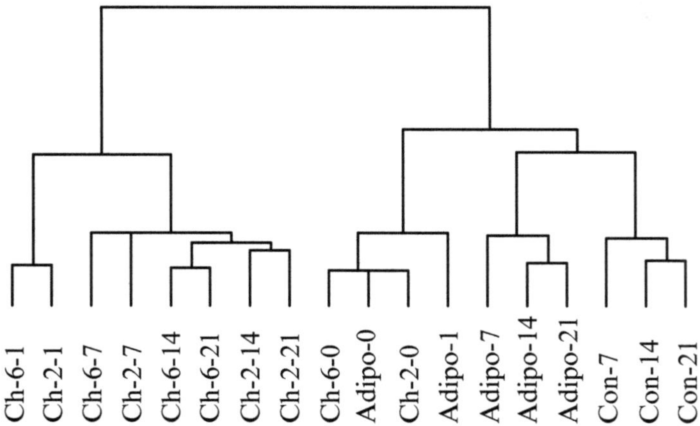

Fig. 10.3 Hierarchial clustering of the samples. Human MSCs were differentiated under three conditions: (**a**) culture as micropellets in a serum-free medium containing BMP-6 and other components previously shown to promote chondrogenic differentiation (Ch-6 experiment); (**b**) culture as micropellets under the same conditions except that the BMP-6 was replaced by BMP-2 (Ch-2 experiment); (**c**) cultures in which adherent log phase cells were transferred to adipogenic medium (Adipo experiment). Control samples were cultured under the same conditions as the Adipo experiment except the medium was complete culture medium (CCM) throughout. Differentially expressed genes from three experiments (Ch-6, Ch-2, and Adipo) and a control experiment with MSCs (Con) were used to generate this figure in dChip. As expected, the Day 0 samples from all three experiments clustered together. In the two chondrogenesis experiments, data from the Day 1 and Day 7 clustered together but the data from Day 14 and Day 21 appeared in separate clusters. Therefore the results indicated that there were differences in the final stages of differentiation with substitution of BMP-6 with BMP-2 in Ch-2 experiments. For the Adipo experiment, the data from Day 1 clustered with the merged super-sample of three Day 0 samples. However, the Adipo Day 7, Day 14 and Day 21 data clustered separately from the other samples *(*reproduced with permission from *20)*

picture each row represents a gene and each column a sample (Fig. 10.4). The genes close to each other in the gene clustering tree have a high similarity in their expression values across all samples. Once the genes are clustered, you can select gene expression patterns and study which genes are present in each pattern. An example of hierarchical clustering of the genes is presented in Fig. 10.4 (reproduced with permission). Also, patterns can be studied for gene functions using GeneOntologies that define the cellular location, molecular function and biological component of the protein product of the gene *(27)*. Enrichment of functions in each pattern can be studied using classification function in dChip that uses hypergeometric distribution to calculate p-values that aid in finding significant GO-terms. An example of GO-term usability is shown in Table 10.1 (reproduced with permission). Also, GO-terms can be searched and used to filter and cluster only genes with a selected function, like transcription factors.

Another useful tool available in dChip is GoSurfer. It can be used to study the GO-term hierarchy and significant differences between GO-terms of two gene lists in a hierarchical tree format. Also genes in a list can be mapped to chromosomes

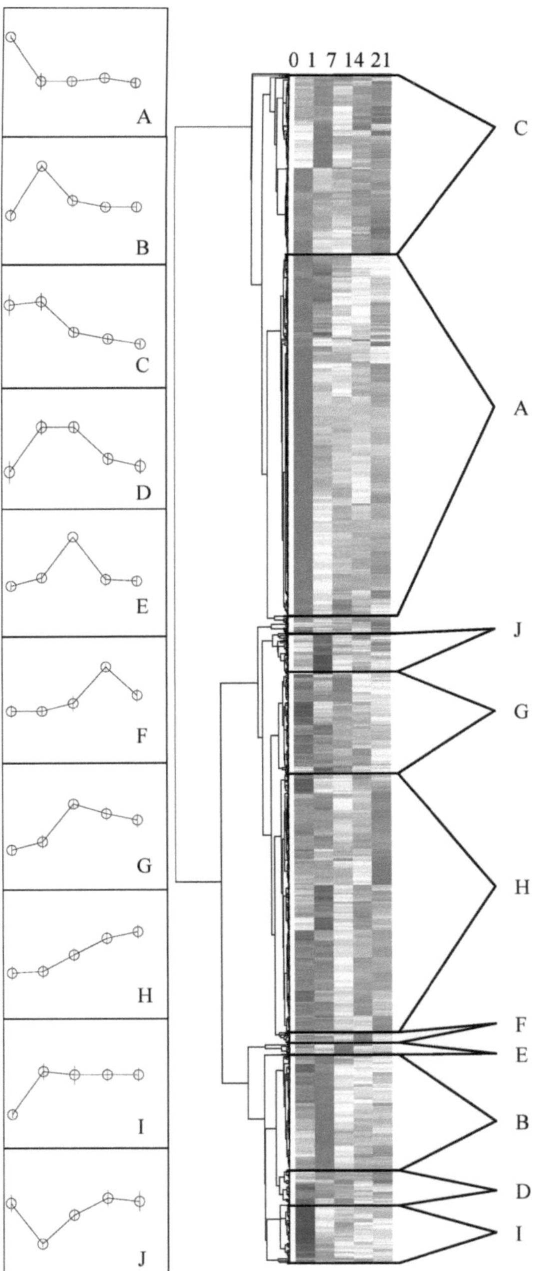

Fig. 10.4 Hierarchial clustering of the genes. Differentially expressed genes (as in Fig. 10.3) for Ch-6 experiment were used to cluster genes hierarchically in dChip program. The red color represents expression level above mean expression of a gene across all samples, the white color represents expression at the mean level, and the blue color represents expression lower than the mean. Co-expressed genes were defined from the clustering picture and average profiles are shown (identified as A-J for Ch-6) on the left of the clustering picture. In the picture a row represents a gene and each column represents a sample from the time course (Day 0, 1, 7, 14, and 21) (reproduced with permission from *20)* (*See Color Plates*)

Table 10.1 Pattern analysis using GeneOntologies. Five most significant GO terms assigned to selected expression profiles (Fig. 10-2) as MSCs differentiated to chondrocytes in the presence of BMP6 (reproduced with permission from *20*)

Pattern (No. of genes annotated)[a]	GO-term (No. of genes)[b]	*p*-value[c]
A (370)	DNA replication and chromosome cycle (21)	$<10^{-6}$
	M phase of mitotic cell cycle (22)	$<10^{-6}$
	mitotic cell cycle (44)	$<10^{-6}$
	nuclear division (21)	$<10^{-6}$
	spindle (10)	$<10^{-6}$
H (245)	skeletal development (25)	$<10^{-6}$
	extracellular matrix (21)	$<10^{-6}$
	organogenesis (42)	0.000010
	morphogenesis (45)	0.000013
	development (55)	0.000120

[a] No. of annotated genes in profile.
[b] No. of genes in profile with specific annotation.
[c] *p*-value <0.01 (based on hypergeometric distribution)

to see if certain sections of chromosomes are more active in some samples than others. We have also used the sample classification by linear discriminant analysis (LDA) function of the dChip to classify samples of unknown classes based on training samples with known classes. In LDA, one must select two classes of samples and use a list of genes obtained, preferably, by comparing the classes to each other. Principal component analysis (PCA) is a useful way to explore the naturally arising sample classes based on an expression profile.

3.10 Pathway Analysis

An important aspect of the microarray data analysis is to find what it means if a certain gene is differentially expressed. Because genes rarely act on themselves, interaction and pathway analysis are a significant part of the analysis. One program we have used for analysis of pathways is GenMAPP 2.0 (Gene Map Annotator and Pathway Profiler), which allows one to visualize gene expression data in a biological context with the graphical and more intuitive format of MAPPs *(28,29)*. A MAPP is a program produced file that graphically shows the biological relationship between gene products. GenMAPP can represent, for example, metabolical pathways, signal transduction cascades, sub-cellular locations, gene families, or lists of genes associated with GO categories. To start the use of GenMAPP, prepare the expression data according to GenMAPP instructions and load it to the program. One of the great features of GenMAPP is that you can assign various colors to many different conditions like up- or downregulation in certain number of samples. We have found it useful to give different shades of red for upregulated genes, either based on magnitude of the change or sample number in which the gene was upregulated. We generally do the same with downregulated genes, using different shades of blue. This way when a MAPP is opened, one can see if the genes in the MAPP are generally up- or downregulated in the experiment. This will enable the finding of activated or silenced pathways in the experiment and will also aid in finding interacting components. In addition, custom MAPPs for hypothesis testing may be drawn with the graphics tools provided in the program.

3.11 Microarray Data Validation

An important part of the microarray experiment is data validation to exclude false positive and false negative results. Microarray data can be validated both on RNA and protein level. RNA level validation methods include RT-PCR and real-time RT-PCR whereas protein level validation can be achieved using protein arrays, enzyme

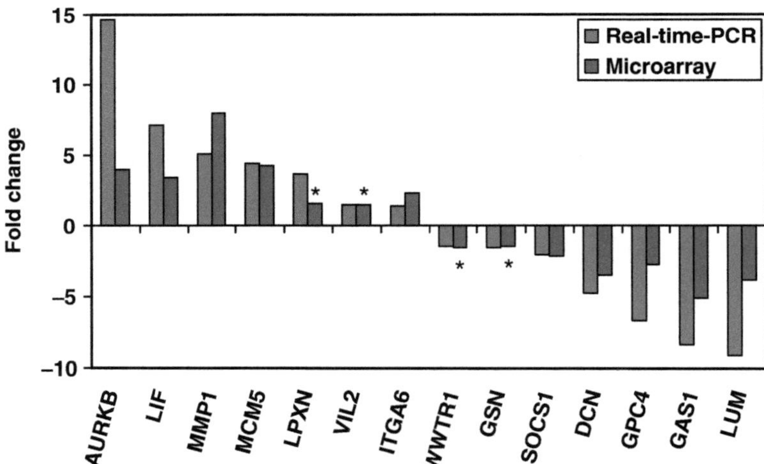

Fig. 10.5 Validation of the microarray data with real-time PCR. MSCs were grown on laser capture microdissection (LCM) slides at 0.5 cells/cm^2. MSCs were fixed and cells from inside (IN) and outside (OUT) of the colonies were isolated with LCM. Total RNA was extracted and used for microarrays with amplification (starting with 10 ng of RNA) and for real-time PCR without amplification (40 ng). The fold changes shown for microarrays are obtained comparing three samples (biological replicates) from outside of the colony to three samples from inside of the colony (mean expression values between groups are compared). Real-time PCR was performed using custom designed low density arrays (LDA, Applied Biosystems) containing 64 genes in triplicate for two samples in a 384 well card. Selected genes that showed significant (based on real-time RT-PCR) up- (positive fold-change) or downregulation (negative fold change) in the outside of the colony compared to the inside of the colony are shown. Asterisk (*) indicates fold changes that were not significant (less than 2-fold change) on microarray (unpublished data) (reproduced with permission from *20*)

linked immunosorbent assay (ELISA), western blotting or immunocytochemistry. Recently developed low density arrays (LDA, Applied Biosystems) utilize the Taqman probe system (Applied biosystems) in a fully customizable 384-well card format with predesigned probes providing a convenient way to quantitate the expression of multiple genes simultaneously. An example comparison of microarray data with real-time RT-PCR is shown in Fig. 10.5 (unpublished data), which shows good correlation between microarrays and real-time RT-PCR. Some of the changes in expression levels detected with real-time RT-PCR were not considered significant on the microarray, demonstrating the higher sensitivity of real-time RT-PCR.

1. To prepare the sample for real-time PCR, convert 10–100 ng of total RNA into cDNA with High-Capacity cDNA Reverse Transcription Kit (Applied Biosystems).
2. Adjust the cDNA samples with DEPC water to a final volume of 200 μL.
3. Mix with equal volume of 2× TaqMan® Universal PCR Master Mix (Applied Biosystems) and transfer the mixture into sample-loading ports on a microfluidics 384-well card, preloaded with TaqMan Gene Expression Assay targets (Applied Biosystems) (*see* **Note 12**).
4. Perform the real-time PCR with ABI PRISM® 7900HT Sequence Detection System (Applied Biosystems) and analyze the data using SDS 2.2 program (Applied Biosystems) with endogenous control for normalization (*see* **Note 13**). Use the following parameters for data analysis: automatic Ct calculation, automatic outlier removal and RQ min/max confidence of 95.0 %. When the baseline RQ min or max values do not overlap with the sample RQ values, the change in expression can generally be considered significant.

4 Notes

1. We have found comparable results with the use of fresh sample, RNAlater or the lysis buffer, but occasionally RNA samples deteriorate after storage at low temperature for many months in spite of the presence of RNAse inhibitors.
2. Instead of spectrophotometry and gel electrophoresis, RNA sample can also be tested by loading it into an RNA LabChip (Caliper Technologies, Mountain View, CA) and analyzing it with Agilent 2100 bioanalyzer (Agilent Technologies, Palo Alto, CA) using the RNA 6000 Nano Assay.
3. After synthesis, the double-stranded cDNA can be stored at −20 °C.
4. We have also used BioArray HighYield RNA Transcription Labeling Kit (Enzo Diagnostics, Farmingdale, NY) successfully.
5. Instead of 16 h incubation, 4 h incubation can be performed by adding 1 μL (200 U) of cloned T7 RNA polymerase (Ambion).
6. Also Sample Cleanup Module (Affymetrix) or Chroma Spin-100 columns (Clontech, Mountain View, CA) followed by ethanol precipitation can be used to clean the generated cRNA.
7. Quality of the cRNA can also be checked with agarose gel electrophoresis or on an Agilent 2100 Bioanalyzer.
8. The array can be stored at 4 °C for up to 3 h if necessary, but must be equilibrated to room temperature before washing and staining.
9. If array is not scanned immediately, it can be placed in the dark at 4 °C, but must be equilibrated to room temperature before scanning.

10. Also GeneArray Scanner (Agilent) can be used.
11. Also various other shareware type softwares like MIDAS, MADAM, MEV and Spotfinder from TIGR (The Institute for Genomic Research), SAM from Stanford and commercially available programs like DecisionSite (Spotfire, Somerville, MA), S-Plus ArrayAnalyzer (Insightful, Seattle, WA), GeneSpring (Silicon Genetics, Redwood City, CA) and PartekPro (Partek, St. Charles, MO) can be used to analyze the microarray data from the generated images or txt-files.
12. Depending on the design (number of genes) of the card, 1–8 samples can be run simultaneously on a same card.
13. Various endogenous controls, like GAPDH, beta-actin and 18 S, can be used for normalization. The expression of the endogenous control cannot change owing to the experimental conditions, and it is good practice to check the expression levels of at least couple endogenous controls before choosing one.

References

1. Schena, M., Shalon, D., Davis, R. W., and Brown, P. O. (1995) Quantitative monitoring of gene expression patterns with a complementary DNA microarray *Science*. **270**, 467–470.
2. Kronick, M.N. (2004) Creation of the whole human genome microarray *Expert Rev. Proteomics*. **1**, 19–28.
3. Bertone, P., and Snyder, M. (2005) Advances in functional protein microarray technology *FEBS J*. **272**, 5400–5411.
4. Giltnane, J. M., and Rimm, D. L. (2004) Technology insight: Identification of biomarkers with tissue microarray technology *Nat.Clin. Pract. Oncol*. **1**, 104–111.
5. Vissers, L. E., Veltman, J. A., van Kessel, A. G., and Brunner, H. G. (2005) Identification of disease genes by whole genome CGH arrays *Hum. Mol. Genet*. **14** R215–223.
6. Lipshutz, R. J., Fodor, S. P., Gingeras, T. R., and Lockhart, D. J. (1999) High density synthetic oligonucleotide arrays *Nat. Genet*. **21**, 20–24.
7. Lockhart, D. J., Dong, H., Byrne, M. C., Follettie, M. T., Gallo, M. V., Chee, M. S., Mittmann, M., Wang, C., Kobayashi, M., Horton, H., and Brown, E. L. (1996) Expression monitoring by hybridization to high-density oligonucleotide arrays *Nat. Biotechnol*. **14**, 1675–1680.
8. BammLer, T., Beyer, R. P., Bhattacharya, S., Boorman,G. A., Boyles, A., Bradford, B.U., Bumgarner, R. E., Bushel, P. R., Chaturvedi, K., Choi, D., Cunningham, M. L., Deng, S., Dressman, H. K., Fannin, R. D., Farin, F. M., Freedman, J. H., Fry, R. C., Harper, A., Humble, M. C., Hurban, P., Kavanagh, T. J., Kaufmann, W. K., Kerr, K. F., Jing, L., Lapidus, J. A., Lasarev, M. R., Li, J., Li, Y. J., Lobenhofer, E. K., Lu, X., Malek, R. L., Milton, S., Nagalla, S. R., O'malley, J. P., Palmer, V. S., Pattee, P., Paules, R. S., Perou, C. M., Phillips, K., Qin, L. X., Qiu, Y., Quigley, S. D., Rodland, M., Rusyn, I., Samson, L. D., Schwartz, D. A., Shi, Y., Shin, J. L., Sieber, S. O., Slifer, S., Speer, M. C., Spencer, P. S., Sproles, D. I., Swenberg, J. A., Suk, W. A., Sullivan, R. C., Tian, R., Tennant, R. W., Todd, S. A., Tucker, C. J., Van Houten B., Weis, B. K., Xuan, S., and Zarbl, H. (2005) Standardizing global gene expression analysis between laboratories and across platforms *Nat Methods* **2**, 351–356.
9. Irizarry, R. A., Warren, D., Spencer, F., Kim, I. F., Biswal, S., Frank, B. C., Gabrielson, E., Garcia, J. G., Geoghegan, J., Germino, G., Griffin, C., Hilmer, S. C., Hoffman, E., Jedlicka, A. E., Kawasaki, E., Martinez-Murillo, F., Morsberger, L., Lee, H., Petersen, D., Quackenbush, J., Scott, A., Wilson, M., Yang, Y., Ye, S. Q., and Yu, W. (2005) Multiple-laboratory comparison of microarray platforms *Nat. Methods*. **2**, 345–350.
10. Larkin, J. E., Frank, B. C., Gavras, H., Sultana, R., and Quackenbush, J. (2005) Independence and reproducibility across microarray platforms *Nat. Methods*. **2**, 337–344.
11. Yauk, C. L., Berndt, M. L., Williams, A., and Douglas, G. R. (2004) Comprehensive comparison of six microarray technologies *Nucleic Acids Res*. **32**, e124.

12. Inomata, N., Tomita, H., Ikezawa, Z., and Saito, H. (2005) Differential gene expression profile between cord blood progenitor-derived and adult progenitor-derived human mast cells *Immunol. Lett.* **98**, 265–271.
13. Robson, P. (2004) The maturing of the human embryonic stem cell transcriptome profile *Trends Biotechnol.* **22**, 609–612.
14. Ivanova, N. B., Dimos, J. T., Schaniel, C., Hackney, J. A., Moore, K. A., and Lemischka, I. R. (2002) A stem cell molecular signature *Science.* **298**, 601–604.
15. Ramalho-Santos, M., Yoon, S., Matsuzaki, Y., Mulligan, R. C., and Melton, D. A. (2002) "Stemness": transcriptional profiling of embryonic and adult stem cells *Science.* **298**, 597–600.
16. Spees, J. L., Olson, S. D., Ylostalo, J., Lynch, P. J., Smith, J., Perry, A., Peister, A., Wang, M. Y., and Prockop, D. J. (2003) Differentiation, cell fusion, and nuclear fusion during ex vivo repair of epithelium by human adult stem cells from bone marrow stroma *Proc. Natl Acad. Sci. U S A.* **100**, 2397–2402.
17. Pochampally, R. R, Smith, J. R., Ylostalo, J., and Prockop, D. J. (2004) Serum deprivation of human marrow stromal cells (hMSCs) selects for a subpopulation of early progenitor cells with enhanced expression of OCT-4 and other embryonic genes *Blood.* **103**, 1647–1652.
18. Sekiya, I., Larson, B. L., Vuoristo, J. T., Cui, J. G., and Prockop, D. J. (2004) Adipogenic differentiation of human adult stem cells from bone marrow stroma (MSCs) *J. Bone Miner. Res.* **19**, 256–264.
19. Sekiya, I., Vuoristo, J. T., Larson, B. L., and Prockop, D. J. (2002) In vitro cartilage formation by human adult stem cells from bone marrow stroma defines the sequence of cellular and molecular events during chondrogenesis *Proc. Natl Acad. Sci. U S A.* **99**, 4397–4402.
20. Ylostalo, J., Smith, J. R., Pochampally, R. R., Matz, R., Sekiya, I., Larson, B. L., Vuoristo, J. T., and Prockop, D. J. (2006) Use of differentiating adult stem cells (marrow stromal cells) to identify new downstream target genes for transcription factors *Stem Cells.* **24**, 642–652.
21. Chen, J. J., Delongchamp, R. R., Tsai, C. A., Hsueh, H. M., Sistare, F., Thompson, K., Desai, V. G., and Fuscoe, J. C. (2004) Analysis of variance components in gene expression data *Bioinformatics.* **20**, 1436–1446.
22. Van Gelder, R. N., von Zastrow, M. E., Yool, A., Dement, W. C., Barchas, J. D., and Eberwine, J. H. (1990) Amplified RNA synthesized from limited quantities of heterogeneous cDNA *Proc. Natl Acad. Sci. U S A.* **87**, 1663–1667.
23. Li, C., and Wong, W. H. (2003) DNA-Chip Analyzer (dChip). In: Parmigiani G, Garrett ES, Irizarry R, Zeger SL, editors. The Analysis of Gene Expression Data: Methods and Software. 1st ed. Springer.
24. Li, C., and Wong, W. H. (2001) Model-based analysis of oligonucleotide arrays: expression index computation and outlier detection *Proc. Natl Acad. Sci. U S A.* **98**, 31–36.
25. Golub, T. R., Slonim, D. K., Tamayo, P., Huard, C., Gaasenbeek, M., Mesirov, J. P., Coller, H., Loh, M. L., Downing, J. R., Caligiuri, M. A., Bloomfield, C. D., and Lander, E. S. (1999) Molecular classification of cancer: class discovery and class prediction by gene expression monitoring *Science.* **286**, 531–537.
26. Eisen, M. B., Spellman, P. T., Brown, P. O., Botstein, D. (1998) Cluster analysis and display of genome-wide expression patterns *Proc. Natl Acad. Sci. U S A.* **95**, 14,863–14,868.
27. Ashburner, M., Ball, C. A., Blake, J. A., Botstein, D., Butler, H., Cherry, J. M., Davis, A. P., Dolinski, K., Dwight, S. S., Eppig, J. T., Harris, M. A., Hill, D. P., Issel-Tarver, L., Kasarskis, A., Lewis, S., Matese, J. C., Richardson, J. E., Ringwald, M., Rubin, G. M., and Sherlock, G. (2000) Gene ontology: tool for the unification of biology. The Gene Ontology Consortium *Nat. Genet.* **25**, 25–29.
28. Doniger, S. W., Salomonis, N., Dahlquist, K. D., Vranizan, K., Lawlor, S. C., and Conklin, B. R. (2000) MAPPFinder: using Gene Ontology and GenMAPP to create a global gene-expression profile from microarray data *Genome Biol.* **4**. R7.
29. Dahlquist, K. D., Salomonis, N., Vranizan, K., Lawlor, S. C., and Conklin, B. R. (2002) GenMAPP, a new tool for viewing and analyzing microarray data on biological pathways *Nat. Genet.* **31**, 19–20.

Chapter 11
Gene Delivery to Mesenchymal Stem Cells

Reza Izadpanah and Bruce A. Bunnell

Abstract Successful gene therapy technology relies on the delivery of the therapeutic product into appropriate target cells. Gene delivery to mesenchymal stem cells (MSCs) has been proposed as a mechanism to promote the augmentation of tissue-engineered replacement systems. In particular, MSCs are attractive targets for gene delivery systems, because they can differentiate, in response to various molecular signals, into many types of committed cells. Introduction of transgene of interest into autologous stem cell types poses an attractive cell-based delivery strategy. MSCs divide rapidly and, because of their high amphotropic receptor levels, are readily transducible with integrating vectors and maintain transgene expression in vitro and in vivo without affecting multipotentiality. The unique biology of MSCs predetermines them to become valuable cytoreagents for gene therapy approaches in future. This chapter describes methods and associated materials for transducing mesenchymal stem cells with a desired nucleic acid.

Keywords Mesenchymal stem cells; gene transfer; viral mediated gene delivery; transfection; transduction; lipofection; plasmid; viral vector.

1 Introduction

Since the identification of mesenchymal stem cells (MSCs), researchers have made great strides in the analysis of the biology of the MSCs for the development of therapeutic applications. Thus, stem cell research promises new type of treatments and possible cures for a variety of debilitating diseases and injuries. MSCs display high degree of stem cell plasticity *(1,2)*. Control of the growth and differentiation of stem cells is a critical tool in the fields of regenerative medicine and tissue engineering. Gene transfer in to MSCs has been done in the effort to increase the efficiency of specified stem cell differentiation. This method is a promising approach for the treatment of many genetic and degenerative diseases *(3)*. The isolation of MSCs from different sources such as bone marrow, adipose tissue, liver, gut, brain allows gene transfer into MSC on engraftments of different organs. Gene transfer into stem cells

From: *Methods in Molecular Biology, vol. 449, Mesenchymal Stem Cells: Methods and Protocols*
Edited by D.J. Prockop, D.G. Phinney, and B.A. Bunnell © Humana Press, Totowa, NJ

may allow stable genetic modification of both the stem cells and of all their progeny cells. It has been shown that gene transfer into hematopoietic cells modulates immune responses, to protect hematopoietic cells against cytotoxic drugs or viral genes, and to restore gene deficiencies owing to either inborn genetic defects or acquired loss of regular gene function *(4–6)*. To deliver a therapeutic gene a carrier molecule called a "vector" must be used. Currently, there are several types of vectors available for the gene transfer including plasmids.

Plasmid vectors are small circular molecules of double stranded DNA that have the ability to efficiently express a mammalian gene from a promoter element. Plasmids have been used extensively for gene delivery and expression. In this approach the desired gene inserted to plasmid, the plasmid is expanded and delivered to a target cell population, in this case MSCs. The entry of plasmids to the MSCs is facilitated by chemical compounds.

Viral vector gene delivery systems are typically replication defective virus particles that have been genetically engineered to deliver specific gene(s) to target cells. Viruses have evolved a way of encapsulating and delivering their genes to human cells in a pathogenic manner. The genome in these viruses has been altered such that the genes involved in disease pathogenesis have been removed and replaced with reporter or therapeutic genes. These viruses are suitable to carry the inserted therapeutic genes. In a viral vector-mediated gene transfer to MSCs, the cells are transduced by the recombinant virus particles. The vector then unloads its genetic material containing the therapeutic gene into the MSCs. The ultimate goal is the production of the therapeutic protein in a cell population that can undergo the efficient engraftment and differentiation in diseased or damaged tissue.

Here we present an overview of gene delivery techniques for use with MSCs. Within this chapter, the significance of gene delivery in stem cells, the types of viral vectors for gene transfer and non-viral vector gene delivery techniques will be discussed.

1.1 Mesenchymal Stem Cells

Adult MSCs can be isolated from many tissues including bone marrow and adipose tissue. These cells could be expanded in vitro over several passages *(7,8)*. MSCs are separated from other cell populations in a tissue sample, including hematopoietic stem cells, based on their adherence to plastic *(9–12)*. Many adult tissues contain populations of stem cells that have the capacity for renewal after trauma, disease, or aging. Bone marrow is the major source of adult MSC that contribute to the regeneration of mesenchymal tissues *(13)*. Bone marrow MSCs are able to differentiate into many different cell lineages including adipocytes, chondrocytes, osteocytes, hepatocytes, myocytes, cardiomyocytes, and neural cells *(14–18)*. MSCs have also been identified in adipose tissue. Adipose tissue MSCs *(19)*, like BMSCs, differentiate *in vitro* toward the osteogenic, adipogenic, neurogenic, myogenic, and chondrogenic lineage when treated with established lineage specific factors *(20–22)*.

1.2 Gene Transfer

1.2.1 Nonviral Gene Delivery

One of the most basic techniques of molecular biology to study a specific protein is expression cloning and expression in cells. Among the gene transfer methods, the non-viral methods are less expensive, easier, and safer to make. They can also be stored for relatively long periods of time without special conditions. However, non-viral gene delivery systems are characterized by lower efficiency of gene delivery due, in part, to endolysosomal degradation *(23)*. Nonviral gene delivery also is limited by transient gene expression as a result of very low integration efficiencies. In this technique, DNA coding for a protein of interest is cloned (using PCR and/or restriction enzymes) into a plasmid known as an expression vector. Plasmid may have special promoter elements to drive production of the protein of interest, and may also have antibiotic resistance markers to help follow the plasmid.

The introduction of genes into mammalian cells in vitro has been accomplished by a variety of methods. Transfection is the uptake of DNA fragments from medium surrounding mammalian cells. Several different transfection techniques are available including chemical gene delivery and liposome transfection. In chemical transfection, chemical components such as calcium phosphate facilitate the transfer of the DNA construct through cell membrane. Liposomes are defined as sacks of lipid into which plasmids are trapped. Liposomes bind and condense DNA spontaneously to form complexes with high affinity to cell membranes. Endocytosis of the complexes followed by disruption of the endosomal membrane appears to be the major mechanism of gene delivery *(23)*.

1.2.2 Viral Mediated Gene Delivery

In order to modify MSCs to treat diseases, the therapeutic gene must be efficiently delivered to the cell in such a way that the gene can be expressed at therapeutic levels and for an extended duration. Several recent reports have described the successful transduction of MSCs by various recombinant virus vector systems (24–31). The recombinant virus systems include oncoretrovirus, lentivirus, adeno-associated virus, and adenovirus derived vectors. These reports have described the successful expression of various transgenes, such as the *Escherichia coli* β-galactosidase and enhanced green fluorescent protein reporter genes. More over, the expression of potentially therapeutic proteins such as Factor VIII, Factor IX, tyrosine hydroxylase, α-L-iduronidase, β-hexosaminadase A, bone morphogenic protein, human growth hormone, TGF-β1, and interleukins 3 and 7. In this chapter, rationale and protocols for the successful production and application of lentivirus vectors for gene delivery to MSCs will be the central focus.

Replication-defective viral vectors can infect a wide variety of cells, often with great specificity, and are very efficient at introducing their own DNA into the host cell. By replacing viral genes with foreign genes of interest, the recombinant viral vectors can effectively transduce cells.

Lentiviruses (LVs), such as HIV-1, are a subfamily of retroviruses that are biologically distinct because they can efficiently infect both proliferating and non-proliferating cells *(32)*. Vector systems derived from LVs may offer a solution to gene delivery in populations of quiescent cells. LV vectors are replication-defective hybrid virus particles composed of core proteins and enzymes of LVs *(gag, pol, env, tat,* and *rev)* and the envelope of an unrelated virus, most typically the vesicular stomatitis virus-G protein (VSV-G). HIV-1-derived vectors have been demonstrated to mediate efficient delivery, integration, and sustained long-term expression of transgenes in post-mitotic cells such as adult neurons and hepatocytes *(32,33)*. Due to disease pathology associated with replication competent HIV, concern exists about the clinical use of HIV-derived vectors. Because of these concerns, self-inactivating (SIN) LV vectors have been constructed. These vectors are safer and have broader applicability as a means of high-level gene transfer and expression in non-dividing cells *(34)*.

The range of cell types that a specific virus can infect (tropism) is determined by the envelope protein found in the lipid bilayer of the virion and the presence of the receptor for this protein on the surface of the target cell. The host range of retroviral vectors can be extended by pseudotyping, where the *env* gene is replaced with that of another virus. The entry of alternative pseudotypes is limited to cells and tissues that express the cellular receptor, except for VSV-G pseudotyped vectors. Pseudotyped LV vectors have been generated with surface glycoproteins from a variety of different enveloped viruses, including Ebola virus, Marburg virus, rabies virus, cat endogenous virus (RD114), Mokola virus, human foamy virus, gibbon ape leukemia virus, amphotropic murine leukemia virus, influenza virus (HA) and respiratory syncitial virus. LV vectors pseudotyped with the RD114 envelope have been shown to successfully transduce human MSCs, but at a lower efficiency than VSV-G pseudotyped vectors *(35)*.

2 Materials

2.1 Cell Culture

1. Nunclon™ cell culture plates (Nunc).
2. Minimal essential medium, alpha medium (α-MEM) with GlutaMax, without ribonucleosides and deoxyribonucleosides (Invitrogen), for mesenchymal stem cell culture.
3. DMEM with 4.5 g/L glucose supplemented with 10% (v/v) fetal bovine serum (D-10) for 293 cell culture.
4. Heat inactivated fetal bovine serum (Atlanta Biologicals).
5. Penicillin/Streptomycin (100 U/mL).
6. 0.25% Trypsin (Invitrogen).

7. Sterile phosphate buffered saline (PBS) pH 7.4 (Invitrogen).
8. Hemocytometer.
9. Trypan blue (Sigma).

2.2 *Transfection*

2.2.1 Calcium Phosphate Transfection

1. 2 × HEPES Buffer (HEPES Buffered Saline): 8.0 g NaCl, 0.37 g KCl, 0.099 g Na_2HPO_4 (MW 141.96), (or 0.125 g Na_2HPO_4. $2H_2O$ (MW 177.96); or 0.188 g Na_2HPO_4. $7H_2O$ (MW 267.96)), 1.0 g, Dextrose, 5.0 g HEPES. Weigh out ingredients, bring to 450 mL with HPLC purified H_2O, adjust pH to exactly 7.1. Bring to 500 mL with sterile distilled H_2O.
2. 2.5 M $CaCl_2$ in 10 m*M* HEPES: 0.238 g HEPES free acid (MW 238.3) (or 0.260 g HEPES sodium salt (MW 260.3), 27.75 g $CaCl_2$ (or 36.75 g $CaCl_2$ ($2H_2O$). Weigh out ingredients, bring to 90 mL with HPLC purified H_2O, adjust pH to 7.2. Fill to 10 mL with sterile distilled H_2O.
3. TE buffer (pH 7.3): 5.0 mL 1 *M* Tris-HCl pH 7.5, 2.0 mL 250 m*M* EDTA pH 8.0; bring to 500 mL with HPLC purified water.
4. Filter all solutions through a 0.2 μm sterile filter and store at −20 °C.

2.2.2 Lipofection

There are many commercially available kits for lipofection. This chapter will present the techniques for two lipofection kits optimized for the transfection of MSCs.

1. Lipofectamine reagent 2 mg/mL (Invitrogen cat# 50470) and Plus reagent 3 mg/mL (Invitrogen cat# 1096402).
2. SuperFect 3 mg/mL (Qiagen cat#1006699).

3 Methods

All transfections are performed in 6-well plates with cells plated 24–48 h before transfection to approximate 2.5 × 10^5 cells per well on the day of transfection. Target concentration of plasmid DNA should be identified according to the size of the construct. (*see* **Note 1**).

A highly pure DNA dissolved in TE should be used in this experiment (*see* **Note 2 and 3**).

3.1 Culture and Maintenance of MSC

In order to maintain proper cell density, MSCs should be passaged before exceeding 80% confluence.

1. When cells reach 75–80% confluence, aspirate media and wash plate 2 × in sterile PBS.
2. Detach cells by adding 0.25% trypsin (2 mL in a 15-cm plate) and incubating for 5 min at 37 °C (*see* **Note 4**).
3. Inactivate trypsin by addition of 8 mL of α-MEM culture media containing 20% heat-inactivated FBS.
4. Transfer cells to sterile 15-mL conical tube.
5. Centrifuge at 2,000 rpm for 4 min to obtain a pellet.
6. Aspirate supernatant and resuspend the pellet in 1 mL PBS.
7. Add 10 μL of cell suspension and 10 mL of Trypan blue to a microfuge tube. (*see* **Note 5**).
8. Add 10 μL of cells/Trypan blue to a hemocytometer and count cells.
9. Replate the desired number of cells to expand the culture. This may be 2×10^5 cells per 35-mm plate or any other number of cells or plate sizes to meet the 60% confluency within 24–48 h.

3.2 Preparation of Lentivirus Vector Plasmids

1. Plasmid DNA is propagated in an *Escherichia coli* host (e.g., DH5α) and purified using standard molecular biology techniques. A suitable volume of sterile Luria Broth (LB) usually 500–1,000 mL is supplemented with 100 mg/mL ampicillin and inoculated with the appropriate *E. coli* recombinant clone. Bacteria are grown to saturation overnight by shaking (200–300 rpm) at 37 °C.
2. The bacteria are pelleted by centrifugation at 2,500 rpm for 45 min at 4 °C.
3. The plasmids are purified using commercially available kits, typically from Qiagen. Plasmids are purified according to manufacturer's instructions using the Endotoxin-free plasmid purification kits (*see* **Note 2**).
4. Once purified, plasmid stocks can be stored frozen or at 4 °C (*see* **Note 6**).

3.3 Transfection for Lentivirus Vector Production

3.3.1 Calcium Phosphate Transfection

The method described here is modified for transfection of MSCs. The calcium phosphate transfection method is a very efficient means of introducing DNA into many cells types including MSCs. According to our experience, routinely 70–90% of MSCs can be efficiently transfected by this method. (*see* **Note 7**).

This method is quite sensitive to the amount of input plasmid. Therefore, the total amount of plasmid DNA should be determined in a separate experiment (*see* **Note 1**).

1. 24–48 h before transfection, the cells should be plated such that they are logarithmically growing on the day of transfection (i.e., 50–60% confluent at the time of transfection). (*see* **Note 8**).
2. Prepare 2 sets of 15-mL tubes.

   Into one set add: 0.5 mL 2×HBS
   Into the other set mix:
   DNA (1–20 μg in TE) to 450 μL of TE
   2.5 *M* $CaCl_2$ in 10 m*M* HEPES pH7.2, 50 μL.

3. Mix DNA and $CaCl_2$ by drawing up and down in a 1,000 μL pipet and add dropwise to the 2×HBS. A fine opalescent precipitate should appear within 10 min (*see* **Note 9**).
4. Let the mixture stand at room temp for 15 min.
5. Aspirate the cell culture media from the culture dish.
6. Wash the culture dish with room temperature PBS 2 times.
7. Add the DNA $CaPO_4$ coprecipitate dropwise to the surface of the media containing the cells. Swirl the plate gently to mix.
8. Incubate the plate overnight.
9. Remove the $CaPO_4$ containing medium on the next day and replace with regular cell culture medium.
10. Transient assays for gene expression in transfected cells are performed 48–92 h posttransfection.

3.3.2 Lipofection

Lipofection or liposome transfection is a technique used to inject genetic material into a cell by means of liposomes. Liposomes are vesicles that can easily fuse with the cell membrane because they are both made of a phospholipid bilayer. Lipofection is usually done using standard kits commercially available.

3.3.2.1 LipofectAMINE PLUS Reagent

(Invitrogen Lipofectamine reagent 2 mg/mL cat# 50470 + Plus reagent 3 mg/mL cat# 1096402).

1. Add 10 μL of the PLUS reagent to appropriate amount of DNA in 50 μL of TE buffer.
2. Mix tube by inverting and allow to complex at room temperature for 15 min.
3. Add 10 μL of LipofectAMINE into 200 μL of normal growth media without serum.

4. Combine the two tubes, gently mix and let sit at room temperature for 15 min.
5. Adjust the total volume to 1,000 µL with sterile purified water.
6. Remove media from the well wash the wells 2 × with PBS.
7. Add the complex from step 5 above on the cells drop wise.
8. Leave to complex for 2 h at 37 °C.
9. After 2 h add 1,000 µL of serum containing media to the well.
10. Leave the cells at 37 °C overnight.
11. On the next day, transfer the transfected cells out of 6-well tissue culture dish to appropriate tissue culture dishes for expansion and/or selection. (*see* **Note 10**).

3.3.2.2 SuperFect Reagent

(Qiagen SuperFect 3 mg/mL cat#1006699)

1. Add 4 µL of SuperFect reagent to appropriate amount of DNA in 50 µL TE buffer.
2. Pipet up and down "gently" 5 times.
3. Leave mixture at room temperature for 10 min.
4. Remove media from the well, add 950 µL of normal cell growth media (including serum).
5. Add the complex drop-wise (swirl the plate to mix).
6. Leave cells to complex for 3 h. (final volume 1 mL).
7. Add 1,000 µL of growth media (final volume 2 mL in each well).

3.3.3 Post Transfection:

1. Replace transfection mixture with regular cell culture media the next day. (*see* **Note 11**).
2. Evaluate the gene and protein expression in transfected cells 48–96 h posttransfection.

3.4 Preparation of Lentivirus Vector Supernatant

There are multiple HIV-based lentiviral vectors available from several different academic laboratories worldwide and some commercial vendors. A discussion on the detailed construction of these vectors is not the focus of this chapter. The majority of LV vectors in use today are termed split-coding vectors. In these vectors, the transfer vector contains only long terminal repeats (LTRs), RNA packaging sequence, Rev-response-element, and transgene driven by an internal promoter. The packaging vector typically has a heterologous promoter (e.g., the cytomegalovirus immediate-early enhance promoter) driving the *gag* and *pol* genes, which give rise to the structural, enzymatic, and regulatory viral genes. In the context of high-titer recombinant virus vector preparations, the accessory genes of HIV (Vif, Vpr, Vpu, and Nef) are dispensable and as such are not contained in the preponderance of lentivirus production plasmids available. Most techniques described in this chapter use VSV-G for envelope, however alternative virus envelopes can be used.

3.4.1 Preparation and Transfection of 293T Cells

293T cells are human embryonic kidney fibroblasts (293 cells) that have stably express SV40 virus large T antigen from an integrated plasmid. They are resistant to G418, but are continuously maintained in the antibiotic. Approximately every 30–45 d, G418 can be added to the culture medium for 12–14 d to select for cells expressing T antigen. The cells are grown in DMEM complete at 37 °C in a water-jacketed incubator in the presence of 5% CO_2. The cells are loosely adherent, and should be split 1:5 every 3–4 days simply by aspirating the spent medium and gently washing the cells off the plate using a small amount of fresh medium (*see* **Note 12**). The 293T cells seem very sensitive to trypsin and EDTA and as such these reagents are not routinely used during passage of the cells. In the first few days after thawing or passaging of the cells, they tend to clump and grow somewhat rounded (but attached) to the culture surface. During protracted culture, the morphology of the cells changes so that they become more flattened and spread out, but even during this period they only adhere loosely to the growth surface. However, the transfection efficiency and viral production of 293Ts will remain quite high.

1. At least 12–18 h before transfection, split the 293T cells into 15-cm dishes (see Note 13). 13×10^6 cells are plated per 15-cm dish. At the time of transfection, cells should be 50-70% confluent.
2. Prepare stock solutions of the 3 plasmids for transfection: the vector plasmid, the packaging plasmid, and the envelope plasmid. The plasmids are transfected in the following stochastic ratios: Vector-5:Packaging-5:Envelope-1. This ratio is typically used when the envelope plasmid expresses VSV-G to reduce the amount of cytopathic effects caused by very high expression levels of VSV-G. When alternative envelopes are used a 1:1:1 plasmid ratio can be used for transfection. For each 15-cm plate, 25 µg of the vector and packaging plasmids and 5 µg of the envelope plasmid (VSV-G) are required. For a typical preparation of 20 plates, 500 µg of plasmid will be required.
3. Mix the necessary volumes of the three plasmids plus purified H_2O to a total volume of 1,125 µL. Spin the DNA solution through a 0.2-m filter (Spin-X, Costar) attached to syringe. If you are doing multiple dishes, simply multiply the volumes of DNA by the number of plates to be done to arrive at the total DNA and solution volume required.
4. Add 125 µL of 2.5 *M* $CaCl_2$ per plate to the DNA solution, for a total volume of 1,250 µL per 15-cm plate.

5. Add an equivalent volume (1,250 μL) of 2×HBS (pH = 7.12) into a 15-mL polystyrene tube. Add the DNA solution **dropwise** to the HBS. After addition of the DNA, gently invert the tube to mix the solutions. DO NOT VORTEX. A fine precipitate should immediately be visible. The solution will become cloudy. Incubate at room temperature for 5 min (*see* **Note 14**).
6. While the DNA precipitate is incubating, carefully aspirate the media on all of the 293T plates. Replace the medium with 15 mL of D-10 to each plate. Make certain to gently distribute the media throughout the plate, as it will just barely cover the surface area.
7. Tilt the dish so that the media collects on one side and slowly add 2,500 μL of DNA precipitate. Immediately swirl the dish to distribute evenly. Incubate in a 5% CO_2 incubator overnight at 37 °C (15–18 h).
8. After 18 h, carefully aspirate the medium from the 293T cells. Gently wash the cells once with sterile PBS. Add 15 mL of D-10 containing 20 m*M* HEPES and 10 m*M* sodium butyrate. Incubate the cells in this medium for 12 h at 37 °C in 5% CO_2. Aspirate the medium, wash the cells once with sterile PBS and refeed with 25 mL of D-10 with 20 m*M* HEPES and culture in the 5% CO_2/37 °C incubator.
9. Collect the medium from the 293T cells in 24 h intervals. Collect in a 250 mL conical tube. Multiple collections can be placed in the same tube. Replace the medium with 25 mL of D-10 with 20 m*M* HEPES and culture in the 5% CO_2/37 °C incubator. Supernatant is collected out to 60 h after the initial transfection. The unconcentrated supernatant can be stored a 4 °C if processed to completion soon thereafter. It can also be frozen at −80 °C and processed later for concentration.

3.4.2 Concentration of Vector Supernatants and Transduction of Targets

Because of the stability of the VSV-G envelope, diluted vector supernatants can be concentrated using several methods. Virus can be concentrated by ultrafiltration, using commercially available units. We routinely use an ultracentrifugation concentration procedure, which is a physical concentration protocol that reduces the volume of the recombinant virus supernatant by 1,000-fold.

1. Combine the media/virus supernatants from all collections into one homogeneous solution, aliquot into 250-mL tubes and spin at 2,000 rpm for 20 min to remove large particles of cellular debris.
2. Filter the supernatant through a 0.45 um filtration unit (Nalgene) to further remove debris and to sterilize the solution. A small aliquot of the filtered supernatant is routinely saved for titer determination (titer before concentration).
3. Clean the buckets and caps of an SW28 rotor with 70% ethanol and let them air dry in a tissue culture hood. Carefully aliquot an equal volume of filtered supernatant to a sterilized polyallomer tube. Each tube/bucket can hold up to 35–38 mL, so the entire rotor can accommodate approx 240 mL total volume. All tubes must be filled to 35 mL to prevent collapse during centrifugation, if you have less than 35 mL in a tube fill the remainder with sterile PBS.

4. Centrifuge the supernatant with the brake on at 19,000 g for 2.5 h at room temperature. Set the brake on the centrifuge to coast to 0 upon deceleration. Remove the tubes from individual buckets using flame-sterilized forceps.
5. Carefully aspirate the supernatant without disturbing the yellowish-white pellet (usually a few mm in diameter) in the center of the tube. Gently resuspend all of the pellets in 1/100th volume of the original starting volume in DMEM base (no additional ingredients) or PBS by scraping and washing with a 1-mL pipet. Rotate the resuspension for 3 h at room temperature by taping the tube(s) to a rotator. The virus solution is mixed by pipeting at hour intervals during this resuspension process.
6. Aliquot the virus suspension into cryovials at convenient volumes, routinely 50–200 μL with one vial of 6 μl for titer determination.

3.4.3 Determination of Biologic Titer: Marker Expression-based Titer

3.4.3.1 Determination of GFP and LacZ Expression

For GFP or LacZ containing vectors, infectious titer is determined by transduction on 293T cells and analyzed by fluorescence microscopy or by X-gal staining. 293 or HeLa cells are plated at 1×10^5 cells/well in 6-well plates 15–20 h before titering. Serial dilutions of the lentivirus vector preparations (0, 1×10^{-1}, 10^{-2}, 10^{-3}, and 10^{-4}) are prepared in D-10 medium and incubated on the cells for a minimum of 4 h or a maximum of 12–15 h (overnight) in 5% CO_2 at 37 °C.

The rationale for the serial dilution of the lentivirus supernatant is to decrease the likelihood of analyzing cells with multiple copies of the vector genome, which would underestimate the number if infectious particles. Upon completion of the transduction, the cells are washed twice in sterile PBS and the media is exchanged for fresh D-10. The cells are expanded in culture for an additional 48 h (4 d total), trypsinized, and collected. The cells are analyzed for the expression of EGFP fluorescence by flow cytometry. Titers are calculated by using the following equation: ([% of EGFP-positive cells/100] dilution factor x number of cells).

3.4.3.2 P24-based ELISA Titer

This is an ELISA based assay to quantitatively measure the HIV-1 P24 antigen level in the lentiviral vector preps. We use a p24 titer assay, which is an ELISA against the p24 viral coat protein. The p24 assay gives a titering range of 1×10^5 to 1×10^6 (transduction units) TU/mL.

3.4.3.3 In Vivo Titration of Lentiviral Vector

To determine the titer of the HIV stocks for in vivo delivery of the lentivirus, serial dilutions of the vector stocks are made and infected into 5×10^5 293 or HeLa cells in a 6-well plate. HIV p24 Gag antigen is measured by ELISA (Alliance; Dupont-NEN or Beckman Coulter) following serial dilution of the concentrated virus. The transduction efficiency for each individual lentiviral preparation is calculated as TU/pg p24. Vector batches are tested for the absence of replication competent lentivirus by monitoring p24 antigen expression in the culture medium of transduced SupT1 lymphocytes for 3 wk. In all cases tested, p24 antigen was undetectable (detection limit, 3 pg/mL).

3.4.4 Detection of Replication-Competent Lentivirus

The greatest potential problem that may be encountered in working with recombinant lentivirus vectors is the formation of a replication-competent lentivirus (RCL) *(36)*. Improvements in lentiviral vector design have been aimed at reducing the probability of generating RCL, which can arise in the producer cell by a recombination event. A recombinogenic event can occur at the level of the plasmid DNA, during the reverse transcription steps of the viral cDNA synthesis, or perhaps even postintegration during chromosome recombination. Although these events are unlikely to produce RCL in any given cell, because of the potentially large numbers of cells involved in virus production, it is necessary to demonstrate the absence of RCL. A variety of assays designed for the detection of RCL exist, but vary in their sensitivity.

The most sensitive method, which is routinely used by our group, was initially developed to detect replication-competent retrovirus in Murine leukemia virus (MLV) preparations. It is denoted as the marker-rescue assay. In this assay, an indicator-cell line is used that has a defective HIV carrying a transgene marker, such as EGFP. It is essential that this cell line have no detectable levels of RCL or the assay will be invalid. The indicator-cell line is transduced with the test lentivirus supernatant and then passaged for a few weeks in vitro. Tissue culture supernatant from this transduced indicator cell line is harvested and tested for its ability to transduce a naïve-cell with the indicator marker. Rescue of the transgene marker, as indicated by transfer and expression of the EGFP gene in the naïve cells, would be indicative of the presence of RCL.

A simpler method to determine whether RCL is present is by transducing a target adherent cell line to high efficiency (>95%), and then passaging the cells for several generations, typically over a 14-21 day period. Supernatant from the transduced targets is then collected and serially tested for its ability to infect naïve cells (depending upon the nature of the transfer vector), but that number should not increase over time. This method assumes that the RCL has maintained the transgene, which may not be correct. An alternative method for RCL detection is to transduce targets to high efficiency and then serially determine capsid (CA or p24) levels in the cell-culture supernatant of the transduced targets. This can be accomplished by using a commercially available ELISA kit (Coulter). CA levels should fall exponentially over time to undetectable levels. The assumption here is that RCL will produce CA, and spreading infection by an RCL will eventually result in high levels of CA in the tissue-culture supernatant.

4 Notes

1. To find the best concentration of DNA a series of transfection is recommended. The best ratios are 0.5, 1, 1.5, 2, and 2.5 μg DNA per transfection.
2. Plasmid DNA for transfection is routinely prepared by ion-exchange chromatography using any of a number of commercial kits (e.g., Qiagen). Using the endotoxin-free version of these kits results in high titer virus, typically in the range of 10^9 (infectious particles) i.p./mL after concentration. The removal of endotoxin is critical as high levels can impede the production of virus particles by killing off of the 293T cells.
3. Optimization of transfection reagent is necessary to avoid any cell toxicity. This reaction should be done in absence of DNA by applying transfection reagents on the cells in different concentration.
4. Prolonged exposure to trypsin is cytotoxic. It is recommended that after 5 min the cells should be examined under a microscope to confirm detachment. Cells that don't appear to be detaching can be coaxed by gently tapping the plate on the lab bench and incubating an additional 2–3 min.
5. Trypan blue provides an indicator of cell viability. Viable cells exclude Trypan blue and dead cells retain it.
6. Plasmids can be stored at 4 °C for short periods of time and if they will be used frequently. Plasmid DNA can be stored at −20 °C for extended periods of time.
7. The critical factor in this method of transfection is pH of solutions. The pH should be checked with sterile pH sticks or reliable pH meters. Since the solutions are of low ionic strength, often 2 or 3 pH meters must be consulted to correctly determine the pH.
8. Transfection of cells older that 48 hours in culture will reduce the efficiency of the DNA uptake by the cells.
9. To save time and DNA we recommend ensuring that the pH of the solutions will allow the formation of a precipitate in the absence of DNA.
10. This transfer is critical as toxic reagents bind to plastic during the transfection process.
11. This change is necessary to avoid cell toxicity owing to transfection reagents.
12. It is important to maintain the 293T cells at subconfluence (never more than 80% confluent). If the cells become confluent, they undergo contact-inhibition and enter stationary phase. 293T cells can be passaged 1:10 every 3–4 d to maintain good growth kinetics. 293T cells used for virus production should be replaced by thawing a new vial of cells after 6 mo of continuous culture, as titers drop off significantly as cells undergo protracted culture.
13. The transfection procedure described here is for 15-cm plates. The transfection can be scaled down to 6-well (35-mm) format and 10-cm or scaled up to much larger surface area vessels. For our vector production, 20–30 of the 15-cm plates are regularly transfected in groups of 10.
14. The calcium phosphate precipitate is made fresh for each round of transfection and it should be added to the cells in a timely manner. The quality of precipitate is crucial; if it is too fine it will not settle upon the cells, and if it is too clumpy the cells cannot take up the particles by endocytosis. The central problem leading to a poor precipitate is the HEPES Buffer must be at the correct pH, if it stays too far in either direction the precipitate will not form efficiently.

References

1. Krause, D. S., et al., (2001) Multi-organ, multi-lineage engraftment by a single bone marrow-derived stem cell. *Cell*, **105**(3): 369–377.
2. Morizono, K., et al., (2003) Multilineage cells from adipose tissue as gene delivery vehicles. *Hum. Gene Ther.* **14**(1):59–66.

3. Hamada, H., et al., (2005) Mesenchymal stem cells (MSC) as therapeutic cytoreagents for gene therapy. *Cancer Sci.* **96**(3): 149–156.
4. Scherr, M. and Eder, M., (2002) Gene transfer into hematopoietic stem cells using lentiviral vectors. *Curr. Gene Ther.* **2**(1): 45–55.
5. Davis, B. M., Humeau, L., and Dropulic, B. (2004) In vivo selection for human and murine hematopoietic cells transduced with a therapeutic MGMT lentiviral vector that inhibits HIV replication. *Mol. Ther.* **9**(2): 160–172.
6. Haviernik, P. and Bunting, K. D. (2004) Safety concerns related to hematopoietic stem cell gene transfer using retroviral vectors. *Curr. Gene Ther.* **4**(3): 263–276.
7. Colter, D. C., et al., (2000) Rapid expansion of recycling stem cells in cultures of plastic-adherent cells from human bone marrow. *Proc. Natl Acad. Sci. U S A.* **97**(7): 3213–3218.
8. Izadpanah, R., et al., (2005) Characterization of multipotent mesenchymal stem cells from the bone marrow of rhesus macaques. *Stem Cells Dev.* **14**(4): 440–451.
9. Friedenstein, A. J., (1976) Precursor cells of mechanocytes. *Int. Rev. Cytol.* **47**: 327–359.
10. Howlett, C. R., et al., (1986) Mineralization in in vitro cultures of rabbit marrow stromal cells. *Clin. Orthop.* 213: 251–263.
11. Lackner, A. A., et al., (1988) Distribution of a macaque immunosuppressive type D retrovirus in neural, lymphoid, and salivary tissues. *J. Virol.* **62**(6): 2134–2142.
12. Kanazawa, T., et al., (2001) Gamma-rays enhance rAAV-mediated transgene expression and cytocidal effect of AAV-HSVtk/ganciclovir on cancer cells. *Cancer Gene Ther.* **8**(2): 99–106.
13. Pittenger, M. F., et al., (1999) Multilineage potential of adult human mesenchymal stem cells. *Science.* **284**(5411): 143–147.
14. Mezey, E., et al., (2000) Turning blood into brain: cells bearing neuronal antigens generated in vivo from bone marrow. *Science.* **290**(5497): 1779–1782.
15. Orlic, D., (2003) Adult bone marrow stem cells regenerate myocardium in ischemic heart disease. *Ann. N Y Acad. Sci.* **996**: 152–157.
16. Ferrari, G., et al., (1998) Muscle regeneration by bone marrow-derived myogenic progenitors. *Science.* **279**(5356): 1528–1530.
17. Theise, N. D., et al., (2000) Liver from bone marrow in humans. *Hepatology.* **32**(1): 11–16.
18. Eglitis, M. A. and Mezey, E. (1997) Hematopoietic cells differentiate into both microglia and macroglia in the brains of adult mice. *Proc. Natl Acad. Sci. U S A.* **94**(8): 4080–4085.
19. Kascsak, R. J., et al., (1982) Virological studies in amyotrophic lateral sclerosis. Muscle Nerve. **5**(2): 93–101.
20. Zuk, P.A., et al., (2001) Multilineage cells from human adipose tissue: implications for cell-based therapies. *Tissue Eng.* **7**(2): 211–228.
21. Zuk, P. A., et al., (2002) Human adipose tissue is a source of multipotent stem cells. *Mol. Biol. Cell.* **13**(12): 4279–4295.
22. Mizuno, H., et al., (2002) Myogenic differentiation by human processed lipoaspirate cells. Plast. Reconstr. Surg. 109(1): 199–209; discussion 210–211.
23. Abdallah, B., Sachs, L., and Demeneix, B. A. (1995) Non-viral gene transfer: applications in developmental biology and gene therapy. *Biol. Cell.* **85**(1): 1–7.
24. Bosch, P., Pratt, S. L. and Stice, S. L. Isolation, characterization, gene modification and nuclear reprogramming of porcine mesenchymal stem cells. Biol Reprod. 2005.
25. Conget, P. A., Allers, C. and Minguell, J.J. (2001) Identification of a discrete population of human bone marrow-derived mesenchymal cells exhibiting properties of uncommitted progenitors. *J. Hematother. Stem Cell Res.* **10**(6): 749–758.
26. Kumar, S., et al., (2004) Osteogenic differentiation of recombinant adeno-associated virus 2-transduced murine mesenchymal stem cells and development of an immunocompetent mouse model for ex vivo osteoporosis gene therapy. *Hum. Gene Ther.* **15**(12): 1197–1206.
27. Lee, K., et al., (2001) Human mesenchymal stem cells maintain transgene expression during expansion and differentiation. *Mol. Ther.* **3**(6): 857–866.
28. Lu, L., et al., (2005) Therapeutic benefit of TH-engineered mesenchymal stem cells for Parkinson's disease. *Brain Res. Brain Res. Protoc.* **15**(1): 46–51.

29. Lu, S., et al., (2005) (Construction of an adeno-associated virus vector expressing CTLA-4Ig and its expression in the transplanted liver allografts). *Zhonghua Gan Zang Bing Za Zhi*, **13**(3): 183–186.
30. Van Damme, A., et al., (2004) Onco-retroviral and lentiviral vector-based gene therapy for hemophilia: preclinical studies. *Semin. Thromb. Hemost.* **30**(2): 185–195.
31. Zhang, X.Y., La Russa, V.F. and Reiser, J. (2004) Transduction of bone-marrow-derived mesenchymal stem cells by using lentivirus vectors pseudotyped with modified RD114 envelope glycoproteins. *J. Virol.* **78**(3): 1219–1229.
32. Naldini, L., (1998) Lentiviruses as gene transfer agents for delivery to non-dividing cells. *Curr. Opin. Biotechnol.* **9**(5): 457–463.
33. Naldini, L., et al., (1996) In vivo gene delivery and stable transduction of nondividing cells by a lentiviral vector. *Science*. **272**(5259): 263–267.
34. Miyoshi, H., et al., (1998) Development of a self-inactivating lentivirus vector. *J Virol*. **72**(10): 8150–8157.
35. Zhang, B., et al., (2004) The significance of controlled conditions in lentiviral vector titration and in the use of multiplicity of infection (MOI) for predicting gene transfer events. *Genet. Vaccines Ther*. **2**(1): 6.
36. Skipski, V.P., Smolowe, A.F. and Barclay, M. (1967) Separation of neutral glycosphingolipids and sulfatides by thin-layer chromatography. *J. Lipid Res*. **8**(4): 295–299.

Part III
Isolation and Characterization of Murine Multipotential Stromal Cells from Bone Marrow

Chapter 12
Isolation of Mesenchymal Stem Cells from Murine Bone Marrow by Immunodepletion

Donald G. Phinney

Abstract Mesenchymal stem cells are typically enriched from bone marrow via their preferential attachment to tissue culture plastic. However, this isolation method has proven ineffective for murine MSCs because various hematopoietic cell lineages survive and/or proliferate on stromal layers in the absence of exogenous cytokines and therefore constitute a large percentage of the plastic adherent population. Although various methods have been described to remove contaminating hematopoietic populations from these cultures none have gained widespread acceptance. Consequently, we developed a method based on immunodepletion to fractionate hematopoietic and endothelial lineages from plastic adherent fibroblastoid (stromal) cells elaborated from murine bone marrow. Colony-forming assays, immunostaining and flow cytometry has been used to validate the effectiveness of this method. Moreover, immunodepleted populations have been shown to exhibit the capacity for multilineage differentiation in vitro and in vivo and therefore retain the characteristics of MSCs. Most recently, we also catalogued the transcriptome of immunodepleted populations via serial analysis of gene expression. Therefore, our immunodepletion scheme provides a means to enrich from murine bone marrow MSCs that's molecular and biological characteristics are well described. Importantly, this immunodepletion method does not employ long-term expansion of plastic adherent cells ex vivo, thereby avoiding the generation of immortalized cell lines.

Keywords Mesenchymal stem cells; marrow stromal cells; immunodepletion; adipogenesis; chondrogenesis; osteogenesis.

1 Introduction

Alexander Friedenstein and coworkers first demonstrated over 40 years ago that transplantation of bone marrow cells under the kidney capsule of mice generated heterotopic osseous tissue that was self-maintaining and self-renewing. Friedenstein went on to show that this activity segregated with the fibroblastoid (stromal) cell

From: *Methods in Molecular Biology, vol. 449, Mesenchymal Stem Cells: Methods and Protocols*
Edited by D.J. Prockop, D.G. Phinney, and B.A. Bunnell

fraction of marrow isolated by preferential adherence to tissue culture plastic (for review *see 1*). Subsequently, various labs verified and extended Friedenstein's work, demonstrating that clonegenic populations of marrow stromal cells enriched by plastic adherence were capable of multi-lineage differentiation in vitro *(2–4)* and in vivo *(5)*. These and other studies validated, in part, the mesengenic process first postulated by Caplan *(6)*, which depicted the step wise differentiation of a putative mesenchymal stem cell (MSC) resident in bone marrow into progressively more restricted precursors of connective tissue cell lineages.

In the absence of definitive antigenic determinants specific to the MSC, most laboratories still enrich these cells from marrow by isolating the plastic-adherent (stromal) cell fraction. This approach provides significant enrichment of MSCs from most species owing to the fact that contaminating hematopoietic elements fail to thrive in these culture conditions and can be eliminated by repeated media changes. However, this enrichment method has proven ineffective for murine MSCs owing to the fact that various hematopoietic cell lineages from murine bone marrow readily adhere to tissue culture plastic, stromal cells or the matrix molecules they secrete *(7–10)*. Furthermore, the plastic adherent stromal fraction of mouse bone marrow supports granulopoiesis and B-cell lymphopoiesis in vitro in the absence of exogenous growth factors and cytokines *(11)*. Therefore, hematopoietic contaminants including long-term repopulating hematopoietic stem cells persist in these cultures even after serial passage *(11,12)*. This fact is best illustrated by studies showing that plastic adherent cells are as effective as whole bone marrow in reconstituting the hematopoietic system of lethally irradiated mice *(7,13–15)*. Therefore, outcome measurements that employ transplantation of plastic adherent cultures to evaluate the in vivo activity of MSCs are confounded by hematopoietic cell engraftment.

Several methods to fractionate hematopoietic and stromal elements from plastic adherent murine bone marrow cultures have been described but none have gained widespread acceptance owing to limitations that adversely affect the yield or quality of cells *(16–18)*. To develop improved methods to enrich MSCs from murine bone marrow, we conducted a detailed analysis of the composition and growth kinetics of plastic adherent murine bone marrow cultures *(12)*. Based on these results we devised an immunodepletion scheme to remove contaminating hematopoietic and endothelial cells from the stromal cell fraction *(19,20)*. This method, described in detail below, yields a cell population that lacks expression of hematopoietic and endothelial markers including CD11b, CD31, CD34, CD45, CD90, and CD117 and expresses markers typical of MSCs, such as CD9, CD29, CD44, CD81, CD106, and Sca1 *(20)*. Immunodepleted populations also retain the capacity to differentiate into adipocytes, chondrocytes, myoblasts and osteoblasts in vitro (Fig. 12.1) and form heterotopic ossicles in vivo *(20)*. Based on their phenotype and differentiation potential, we refer to these cells as immunodepleted murine MSCs (IDmMSCs). Most recently, we also catalogued the transcriptome of IDmMSCs using serial analysis of gene expression *(21)*. Therefore, the molecular phenotype, differentiation potential and biological function of these populations are well described.

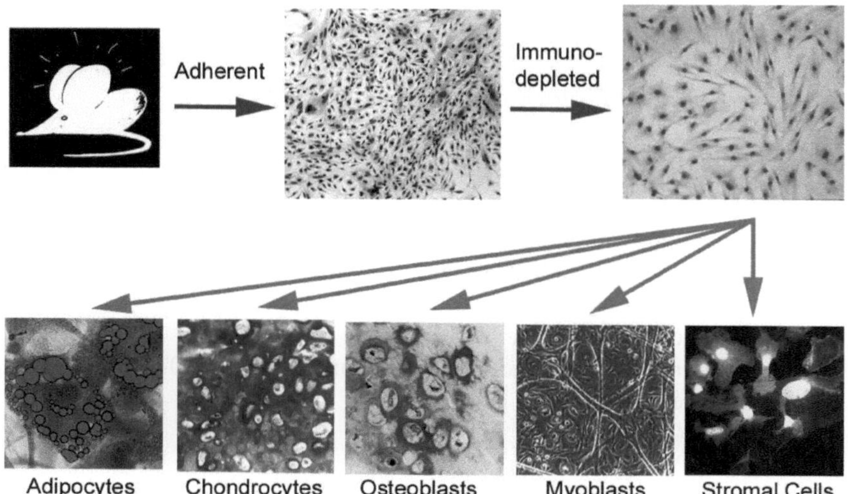

Fig. 12.1 Purification and differentiation of IDmMSCs. The top panel illustrates how immunodepletion removes contaminating hematopoietic lineages from plastic adherent cultures elaborated from murine bone marrow. Images are Geimsa stained plastic adherent populations before and following immunodepletion. The bottom panel shows the potential of IDmMSCs to differentiate into chondrocytes, adipocytes, osteoblasts, myoblasts and hematopoiesis-supporting stroma when cultured under the appropriate conditions in vitro (*See Color Plates*)

Notably, several recent publications have described "novel" methods to isolate murine MSCs by culturing plastic adherent marrow cells long-term and then selecting for "stem cells" by plating at limiting dilution *(22–24)*. Using this approach a small fraction of cells can be isolated that exhibit rapid cell growth (not a requisite characteristic of stem cells) and a multi-potent phenotype. However, it is well established that restricted expression of telomerase in somatic cells is not used as a tumor suppressor mechanism in mice. Therefore, most murine cells exhibit a high frequency of immortalization in vitro by virtue of their telomerase activity *(25)*. Consequently, these "novel" methods for stem cell isolation in actuality are identical to those used to establish hundreds of murine marrow stromal cell lines during the past few decades *(9)*. As expected, "stem cells" isolated by long-term propagation/selection share many traits with marrow stromal cell lines including variability in their capacity to support hematopoiesis, aberrant expression of hematopoietic and endothelial lineage markers such as CD34, CD45, Mac-1, Thy-1, CD8, B220, and factor VIII, a strong tendency to differentiate along the adipogenic lineage, loss of chondrogenic potential, a high frequency of chromosomal abnormalities and lack of cellular senescence. Recently completed comparative studies by our laboratory have shown that IDmMSCs are distinct with regard to phenotype and function as compared to marrow stromal cell lines or populations isolated by long-term propagation in vitro *(21)*.

2 Materials

2.1 Isolation and Culture of Murine Bone Marrow

1. 10-cc syringes with 22-gauge needles, a 30-cc syringe with an 18-gauge needle, surgical forceps (straight and curved), surgical scissors.
2. Dulbecco's Minimum Essential Medium, alpha modification (α-MEM) with L-glutamine and without ribonucleosides and ribonucleotides (Gibco Invitrogen Corp.) (*see* **Note 1**).
3. Fetal bovine serum (FBS) (*see* **Note 2**).
4. Harvest buffer: Hanks balanced salt solution (HBSS), 100 U/mL penicillin, 100 μg/mL streptomycin.
5. Cell strainers (70 μm).
6. Tissue culture dishes (100 mm).
7. Complete culture medium: α-MEM, 10 % FBS, 100 U/mL penicillin, 100 μg/mL streptomycin.

2.2 Harvesting Plastic Adherent Marrow Cells

1. Trypsin-EDTA (0.25% Trypsin).
2. Teflon cell scrapers.
3. Serum-free medium: α-MEM, 100 U/mL penicillin, 100 μg/mL streptomycin.
4. Complete culture medium.
5. Rotator at 4 °C.

2.3 Preparation of Antibody-Conjugated Dynabeads®

1. Dynabeads® M-280 Streptavidin or CELLection™ Biotin Binder Kit. (Dynal Biotech).
2. Dynal MPC®-S magnetic particle concentrator (Dynal Biotech).
3. Biotinylated rat anti-mouse CD11b (Cat. No. 553309), biotinylated rat anti-mouse CD34 (Cat. No. 553732), biotinylated rat anti-mouse CD45 (Cat. No. 553078) antibodies (BD Biosciences Pharmingen).
4. Antibody diluent: HBSS, 0.1% BSA.
5. PBS.
6. Complete culture medium.

2.4 Immunodepletion

1. Dynal MPC®-S magnetic particle concentrator.
2. Anti-CD11b, anti-CD34 and anti-CD45 conjugated Dynabeads®.

3. HBSS.
4. Trypan blue (0.25% solution).
5. T-75 culture flasks.

2.5 *Phenotypic Characterization of IDmMSCs*

1. Wash buffer: HBSS, 1% BSA, 0.025% sodium azide.
2. Trypsin-EDTA (0.25% solution).
3. Paraformaldehyde (2%).
4. Methanol (ice-cold).
5. Mouse Fc-block™, anti-mouse CD16/CD32 monoclonal antibody (BD Pharmingen).
6. Fluorescent-conjugated primary and/or secondary antibodies of interest.

2.6 *Differentiation*

2.6.1 Adipogenic Differentiation

1. 6-well plates or 35-mm tissue culture dishes.
2. Working solution of oil red O (*see* **Note 3**).
3. Toluidine blue (1% solution).
4. Methanol (ice-cold).
5. Adipogenic induction medium: α-MEM, 10^{-8}*M* dexamethasone, 20 μ*M* ETYA (5,8,11,14-Eicosatetraynoic acid), 0, 25 μg/mL insulin, 10% rabbit serum.
6. Adipogenic maintenance medium: α-MEM, 20 μ*M* ETYA, 25 μg/mL insulin, 15% rabbit serum.

2.6.2 Chondrogenic Differentiation

1. ITS-plus premix: 6.25 μg/mL bovine insulin, 6.25 μg/mL transferrin, 6.25 μg/mL selenous acid, 5.33 μg/mL linoleic acid, and 1.25 mg/mL bovine serum albumin (Gibco Invitrogen Corp.).
2. Chondrogenic induction medium: high glucose DMEM, 10 ng/mL TGF-β3, 100 nM dexamethasone, 50 μg/mL ascorbic acid-2-phosphate, 100 n*M* sodium pyruvate, 40 μg/mL proline, ITS-plus premix.
3. Hypertrophic medium: high glucose DMEM, 1 n*M* dexamethasone, 50 μg/mL ascorbic acid-2-phosphate, 100 n*M* sodium pyruvate, 40 μg/mL proline, 20 n*M* β-glycerol phosphate, 50 ng/mL thyroxine, ITS-plus premix.
4. Fixative solution: 20% formaldehyde, 1X PBS.
5. Toluidine blue (1% solution).
6. Sodium borate (1% solution).

7. Acetic acid (1% solution).
8. Light green (1% solution).
9. Safranin O (0.1% solution).

2.6.3 Osteogenic Differentiation

1. 2% solution of Alizarin Red S (*see* **Note 4**).
2. Osteo-inductive medium: high glucose DMEM, 10 % FBS, 10 m*M* β-glycerol phosphate, 50 μg/mL ascorbic acid, 10^{-8}*M* dexamethasone.

3 Methods

3.1 Isolation and Culture of Murine Bone Marrow

1. Typically, bone marrow is harvested from the long bones of 40 male, FVB/n mice at 3–4 wk of age. The long bones of mice at this age are relatively pliable and less likely to splinter during dissection. Also, the cellularity of the bone marrow is greater as compared to older mice thereby producing higher overall cell yields.
2. To isolate marrow, mice are euthanized by exposure to carbon dioxide gas. The animal carcass is then rinsed liberally with 70% ethanol, an incision is made around the perimeter of the hind limbs were they attach to the trunk and the skin is removed by pulling toward the foot, which is cut at the ankle bone. This eliminates further contact of the hind limb with the animal's fur, which is a source of contaminating bacteria. The hind limbs are then dissected from the body trunk by cutting along the spinal cord using care not to damage the femur. Limbs are stored on ice in HBSS supplemented with 1 × penicillin/streptomycin while awaiting further dissection.
3. Further dissection of the hind limbs is done in a sterile cabinet. Each hind limb is bisected by cutting through the knee joint and the connective tissue is removed from both the tibia/fibula and the femur. In the former case, it is easiest to pull the attached muscle and connective tissue at the ankle toward the growth plate, which can then be easily removed in one piece by gently prying off the growth plate. Removing muscle and connective tissue from the femur is best accomplished by scraping the diaphysis of the bone clean then pulling the tissue toward the ends of the bone. Once again, detachment of the growth plate and the ball joint at each end of the bone aides in removal of the connective tissue. After cleaning, the bones are stored in harvest buffer on ice in a 50-mL conical tube (*see* **Note 5**).
4. Extrusion of the bone marrow is performed in a standard bio-safety cabinet (BL-2) using proper sterile technique. The ends of the tibia and femur are cut just below the end of the marrow cavity, which is evident by the transition from a red

to white coloration in the bone using a pair of sharp scissors. Be careful not to splinter the bones during the cutting process. A 22-gauge needle attached to a 10-cc syringe containing complete media is then inserted into the spongy bone exposed by removal of the growth plate. The marrow plug is then flushed out the cut end of the bone with 0.5 mL of complete media and collected in a 50-mL conical tube on ice (*see* **Note 6**). After purging is complete the marrow plugs are dissociated into a single cell suspension by repeated passage (3 ×) through an 18-gauge needle attached to a 30-mL syringe. The cell suspension is then filtered through a 70-μm strainer to remove any bone spicules. The yield and viability of cells is determined by Trypan blue exclusion and counting on a hemocytometer. Typically, we obtain approximately 1×10^9 bone marrow cells from 40 donors.

5. Bone marrow cells are diluted in complete medium to a density of 10×10^6 cells/mL and aliquots (8 mL) are distributed into 100-mm culture dishes (1.45×10^6 cell/cm^2). The plates are cultured undisturbed at 37 °C with 5% CO_2 in a humidified chamber. After 72 h the nonadherent cells that accumulate on the surface of the dish are resuspended by gentle swirling, aspirated and replaced with 8 mL of complete medium (*see* **Note 7**).
6. After an additional 4 d of culture the plates are washed with serum-free medium and fed with 8 mL of complete medium. At this stage the cultures typically exhibit 1 of 2 characteristics. First, plates may contain distinct colonies of fibroblastic cells that vary in size and composition with small numbers of hematopoietic cells interspersed between the colonies. Cultures with these characteristics typically produce good yields of MSCs (>15%). Alternatively, plates may contain small colonies of fibroblastic cells that are intermixed within dense patches of cells that take on a "cobblestone" appearance. Cultures with these characteristics typically produce poor yields of MSCs (<5%).
7. Plates are cultured a total of 8-10 days before immunodepletion. Typically, cultures are harvested when distinct fibroblastic colonies greater than 5 mm in diameter are evident on the plates. Plates containing predominantly small colonies or loose aggregates of fibroblastoid cells can be cultured for several additional days to increase yields (*see* **Note 8**).

3.2 Harvesting Plastic-Adherent Marrow Cells

1. Plates are washed with 5 mL of serum free medium and incubated for Approx 5 min at 37 °C with 4 mL of 0.25% trypsin/EDTA. A small amount of FBS (0.5 mL) is then added to inactivate the trypsin. Cells are collected by gentle scraping using a cell scraper (*see* **Note 9**). After scraping each plate is rinsed once with a small amount of complete medium to remove residual cells. Cells are pooled and stored on ice in a 50-mL conical tube.
2. Cells are collected by centrifugation at 500 *g* for 15 min at 4 °C and the cell pellet is resuspended in 20 mL of HBSS. Importantly, resuspend the cell pellet by flicking the bottom of the tube repeatedly using moderate force before adding

media. The pellet contains a large amount of extracellular matrix that will cause the cells to clump together into an insoluble aggregate if media is added directly to it. Formation of cell aggregates will dramatically reduce the total cell yield.
3. Cells are washed 2 × as described above, suspended in 20mL of HBSS and then counted on a hemocytometer. Collect the cells a final time by centrifugation.
4. Each depletion reaction can accommodate up to 40×10^6 cells and is performed in a final volume of 1mL. Therefore, resuspend up to 40×10^6 cells in 1mL of HBSS. If the yield is greater than 40×10^6 cells divide the sample into 2 or more equal aliquots, each suspended in 1mL of HBSS. Transfer each aliquot to a 1.5mL Eppendorf tube and incubate on a rotator for approx 60min at 4°C.
5. Depending on the number of plates to scrape, harvesting of MSCs can take up to 2h.

3.3 *Preparation of Antibody-Conjugated Dynabeads®*

1. We currently use Dynabeads® M-280 Streptavidin super-paramagnetic polystyrene beads to perform the immunodepletion (*see* **Note 10**). These are supplied as a suspension containing 6.7×10^8 Dynabeads® per mL (10mg/mL). One milligram of Dynabeads® is saturated by incubating with 5–10μg of biotinylated antibody assuming 100% of the antibody is biotinylated. We use 5 beads per cell for each immuno-depletion and perform 3 successive rounds using antibodies against CD11b, CD34, and CD45 (*see* **Note 11**).
2. Aliquot the appropriate amount of streptavidin-conjugated Dynabeads® into 3 separate 1.5-mL Eppendorf tubes labeled CD11b, CD34, and CD45. Place the tubes on the magnetic particle concentrator (MPC) for approx 1min. Slowly remove the liquid using a P-200 pipet. Be careful not to aspirate the beads, which may roll down the sides of the tube during removal of the liquid.

Sample Calculation

40×10^6 cells × 5 Dynabeads®/cell = 200×10^6 Dynabeads;
200×10^6 Dynabeads × 1 mL/6.7×10^8 Dynabeads = 298.5μL Dynabeads.

3. Wash the Dynabeads 3 × by removing the tubes from the MPC, suspending the beads in 500μl of PBS, returning the tubes to the MPC and slowly aspirating the liquid.
4. Dilute each biotinylated antibody (5–10μg/mg Dynabeads) to the appropriate concentration in a total volume of 100μL of antibody diluent. Resuspend the Dynabeads (calculated from above) in the respective antibody solution (calculated from the following) and incubate for 30min at 4°C with gentle agitation every 5min.

Sample Calculation

298.5 μL Dynabeads × 10mg\mL = 2.985mg Dynabeads;
2.985 mg Dynabeads × 10μg antibody/mg = 29.85 μg antibody
29.85 mg antibody ÷ 0.5mg\ml = 59.7 μL antibody

5. Place the Eppendorf tubes on the MPC and remove the antibody solution. Wash the beads 7 × as described above with 500 µL of antibody diluent. Store the antibody-conjugated Dynabeads in 500 µL of antibody diluent at 4 °C until ready for use (*see* **Note 12**).
6. Preparation of antibody-conjugated Dynabeads requires approx 45 min.

3.4 Immunodepletion

1. Place the Eppendorf tube containing the anti-CD11b conjugated Dynabeads on the MPC, remove the buffer by aspiration and remove the tube from the MPC. Retrieve the suspension of marrow cells from the rotator and add then to the Eppendorf tube containing the anti-CD11b conjugated Dynabeads. Thoroughly mix the cells and anti-CD11b conjugated Dynabeads using a pipet (P-1000). Incubate the suspension at 4 °C on the rotator for approx 45 min.
2. Repeat the procedure using the anti-CD34 and then the anti-CD45 conjugated Dynabeads as follows. Retrieve the Eppendorf tube containing the anti-CD11b Dynabeads® suspension from the rotator and place in the MPC. Retrieve the anti-CD34 conjugated Dynabeads from the refrigerator, place on the MPC and remove the buffer by aspiration. Transfer the cell suspension from the Eppendorf tube containing the anti-CD11b conjugated Dynabeads to the Eppendorf tube containing the pellet of anti-CD34 conjugated Dynabeads. Mix the latter thoroughly and incubate at 4 °C on a rotator for 45 min. Repeat the procedure using the anti-CD45 conjugated Dynabeads.
3. Transfer the cells suspension to a 50-mL conical tube and dilute to a final volume of 20 mL with HBSS. Collect the cells by centrifugation at 500 *g* for 15 min at 4 °C. Repeat the wash and suspend the cells in 10 mL of HBSS (remember to disperse the pellet by agitation before adding the buffer). Remove an aliquot of cells (10 µL), mix with 15 µL of HBSS and 25 µL of 0.25% Trypan blue and count the number of viable cells on a hemocytometer.
4. Collect the remaining cells by centrifugation, suspend in complete medium and plate approximately 1×10^6 immunodepleted murine MSCs (IDmMSCs) per T-75 flasks. Culture the cells at 37 °C with 5% CO_2 in a humidified chamber with medium changes 2–3 × weekly (*see* **Note 13**).
5. A typical yield of IDmMSCs from FVB/n mice is typically 15–20% of the total number of plastic adherent cells harvested (*see* **Note 14**). Cell viability is always greater than 90%.

3.5 Phenotypic Characterization of IDmMSCs

1. The expression profile of surface antigens on IDmMSCs can be evaluated by flow cytometry. Typically, IDmMSCs are cultured for approx 5 d before analysis to ensure good cell viability during the sorting procedure (*see* **Note 15**).

2. IDmMSCs are harvested by incubation in trypsin-EDTA (0.25%) (*see* **Note 16**). Cells recovered by centrifugation are suspended in wash buffer and counted on a hemocytometer. Cell density is adjusted to 5 $\times 10^5$ cells/mL and aliquots (0.5 mL) are transferred to 1.5 mL Eppendorf tubes and incubated for 30 min on a rotator at 4 °C.
3. IDmMSCs are collected by centrifugation at 500 *g* for 10 min at 4 °C and the wash buffer is removed using a pipet (P-1000). The pellet is suspended in 50 µL of wash buffer containing 0.25 µg per 1 $\times 10^6$ cells of Mouse Fc-block™ (add 0.0625 µg in 50 µL of wash buffer for 2.5 $\times 10^5$ cells) and incubated on ice for 5 min (*see* **Note 17**).
4. Cells are then incubated with the appropriate dilution of primary antibody prepared in 50 µL of wash buffer, which is added directly to the cell suspension. Removal of the Fc-block is not necessary. Also, multiple antibodies can be added to the cells simultaneously provided that their detection systems are compatible. The cell suspension is incubated for 30 min at 4 °C in the dark with gentle agitation every 10 min.
5. Cells are collected by centrifugation and washed 2 × with 500 µL of wash buffer (or 3 × if a biotin-conjugated primary antibody is used). After each wash agitate the cell pellet before addition of fresh buffer or antibody solution.
6. If using a fluorescent-conjugated secondary antibody, dilute it in wash buffer (100 µl), add it to the cell pellet and incubate for 20 min at 4 °C in the dark.
7. Cells are collected by centrifugation and washed 2 × as described above. After the final wash suspend the cell pellet in 0.5–1.0 mL of wash buffer for analysis.
8. If data cannot be acquired immediately after staining, suspend the cell pellet in 250 µL of wash buffer, mix, and then add an equal volume of 2% paraformaldehyde.

Store tubes wrapped in foil at 4 °C overnight and analyze in several days. Remember to use the appropriate isotype controls for each fluorescent-conjugated antibody evaluated.

9. To detect expression of intracellular antigens cells must fixed and then permeabilized before staining. We typically fix cells in paraformaldehyde (2%) and permeabilize them by incubation in ice-cold methanol for 15 min. Cells are then washed several times and incubated with Fc-block followed by the appropriate primary and/or secondary antibodies.
10. IDmMSCs are known to lack expression of CD11b, CD31, CD34, CD45, CD90, and CD117. The cells uniformly express CD9, CD29 and CD81 and to a lesser extent CD44, CD106, and Sca1.

3.6 Differentiation of IDmMSCs Into Connective Tissue Lineages In Vitro

1. IDmMSCs are typically cultured for 5–7 days before exposure to medium formulations that induce differentiation toward connective tissue cell lineages (*see* **Note 15**).

3.6.1 Adipogenic Differentiation of IDmMSCs

1. IDmMSCs are plated in the appropriate tissue culture vessel (we typically use 6- well plates or 35-mm dishes) at a density that achieves 90% confluence and cultured in complete culture medium.
2. The following day the medium is replaced with adipogenic induction medium.
3. After 48 h the medium is replaced with adipogenic maintenance medium and the cells cultured continuously for several weeks with medium changes 2–3 × weekly. Accumulation of fat droplets becomes apparent after 4–7 d (*see* **Note 18**).
4. To visualize adipocytes, cultures are washed with PBS, fixed in ice-cold methanol for 2 min and then stained with a working solution of oil red O for 30 min. Cultures are then rinsed with tap water and counter-stained with Toluidine blue.

3.6.2 Chondrogenic Differentiation of IDmMSCs

1. IDmMSCs are induced to differentiate into chondroctyes by exposing the cells to TGF-β3 in a 3-dimensional culture system (micromass) using procedures initially described by Johnstone et al. *(27)* and MacKay et al. *(28)*.
2. IDmMSCs (500,000 cells/mL) are suspended in an appropriate volume of chondrogenic induction medium and aliquots (5 mL, 2.5 × 10^5 cells) are transferred to 15-mL Falcon tubes. The tubes are centrifuged at 500 *g* for 10 min at 4 °C to pellet the cells (*see* **Note 19**).
3. The Falcon tubes containing the cell pellets are removed from the centrifuge, wiped thoroughly with 70% ethanol, placed in a rack in a tissue culture incubator and incubated at 37 °C in a humidified chamber with 5% CO_2 for 4 wk. Be careful to loosely attach the lids to the tubes so that gas exchange can occur. Medium is changed 2–3 × weekly. Use every precaution so that the cell pellets are not disturbed during medium changes.
4. After 4 wk of continuous culture a significant increase in the size of the micromass pellet should be notable by visual inspection (*see* **Note 20**). Pellets can be harvested for histological analysis at this stage if desired.
5. To induce hypertrophy of the chondrocytes, remove the medium, replace with hypertrophic medium and culture the pellets for an additional 3–4 wk.
6. At the end of the culture period carefully collect the pellets, incubate in fixative solution over night and then process in paraffin or glycol methacrylate (GMA) using standard protocols.
7. Histological sections (3–4 μm) are deparafinized, hydrated and then stained with % Toluidine blue (with or without 1% sodium borate) for 5 min. Sections are then dehydrated, cleared and mounted with a cover slip. Toluidine blue stains proteoglycans a metachromatic red-purple color and nuclei blue. Inclusion of sodium borate intensifies the blue color.
8. Alternatively, deparafinized sections can be stained with Hematoxylin for 7 min, washed in running tap water for 10 min, stained with 1% light green for 4–5 min,

slowly dipped in 1% acetic acid 4 times and stained with 0.1% Safranin O for 8 min. Sections are then dehydrated, cleared and a cover slip is applied. Safranin O stains proteoglycans a yellow-red color and nuclei red (*see* **Note 21**).

9. Microscopic evaluation of the pellets should reveal oat-shaped chondrocytes within lacunae interspersed by a significant amount of extracellular matrix. Pellets can also be stained with an antibody specific for type-II collagen, which is expressed exclusively in cartilage.

3.6.3 Osteogenic Differentiation of IDmMSCs

1. IDmMSCs are plated in the appropriate tissue culture vessel (we typically use 6-well plates or 60-mm dishes) at a density that achieves approximately 60% confluence and cultured in complete culture medium.
2. The following day complete culture medium is replaced with osteo-inductive medium and the cells are cultured continuously for up to 3 wk with medium changes 2–3 × weekly (*see* **Note 22**).
3. To evaluate the extent of mineralization, cell monolayers are washed with PBS, fixed for 10 min in 10% formaldehyde at room temperature and washed with distilled water. The cell monolayer is then stained 5 min with a 2% solution of Alizarin Red S and washed with water.
4. The extent of osteogenic differentiation can be determined by quantifying the amount of Alizarin Red S dye *(29)* or the amount of calcium *(20)* bound to the extracellular matrix as previously described.

4 Notes

1. α-MEM media containing ribonucleosides and ribonucleotides is toxic to IDmMSCs. The reason for this remains indeterminate. Nevertheless, we mistakenly used this media formulation once and found that after approximately one week essentially all of the IDmMSCs in culture were killed.
2. FBS is lot selected to optimize cell growth and limit the extent of cellular differentiation. MSCs are cultured for 2 wk in medium supplemented with 10% FBS from sample lots obtained from the appropriate vendor. The extent of cell proliferation is measured every few days to generate a growth curve. Levels of alkaline phosphatase (ALP), a marker of osteoblast differentiation, are also measured using a commercial kit (Sigma-Aldrich). Lots that stimulate strong cell growth but fail to induce significant ALP activity are selected for further use.
3. A stock solution of oil red O is prepared by dissolving 0.5 g of dye in 100 mL of isopropanol and passing the solution through a 0.22 µm filter. A working solution is prepared by adding 6 mL of stock solution to 4 mL of water. The solution is left at room temperature for 1 h and then passed through a 0.22-µm filter.
4. A working solution (2%) of alizarin red S is prepared by dissolving the dye in distilled water and adjusting the pH to 4.1 with ammonium hydroxide. The solution is then passed through a 0.22 µm filter to remove particulates.
5. Surgical instruments used to dissect the hind limbs are typically replaced with clean, sterile instruments after every 5–10 mice. We generally use a dry bead sterilizer to sterilize instruments between animals.

6. Recently a method was published that employs high speed centrifugation to harvest marrow from the long bones of mice *(26)*. We do not recommend the use of this procedure because it strips off the endosteal and/or periosteal layer of the bone thereby contaminating the cultures with osteoprogenitor cells. Owing to their robust growth in vitro, these cells may quickly become the most prevalent cell type within the plastic adherent cultures.
7. Do not vigorously wash the culture plates at this time. Removal of the majority of hematopoietic cells at this stage will severely curtail the growth of the fibroblastoid cells thereby dramatically reducing the overall yield of IDmMSCs.
8. It is important to familiarize oneself with the morphology of the plastic adherent cultures at different stages of expansion. For example, initially after plating the vast majority of cells in the culture are nonadherent and only small number of stellate-shaped cells is seen attached to the culture dish. After the first medium change many more adherent cells are visible. These cells adopt a spindle-shaped morphology and begin to migrate together to form clusters. After 5–6 d cells within clusters begin to proliferate rapidly forming large fibroblastoid colonies. Some colonies are comprised of a morphologically homogeneous population of fibroblastoid cells whereas others appear heterogeneous and may contain a large number of hematopoietic cell types. Finally, the total culture time of the plastic adherent populations is limited because fibroblastoid cells within the interior of expanding colonies begin to differentiate along the osteoblastic lineage.
9. Based on our experience, one of the most challenging aspects of this procedure for the novice is harvesting adherent cells from dishes before immunodepletion. Unfortunately, exposure of cells to trypsin alone is not sufficient to detach the cells from the dishes (we have explored many chemical/biological alternatives without success). Therefore, a physical method such as scraping must also be employed. Unfortunately, the binding avidity and flattened morphology of the cells makes them susceptible to physical injury and death if too much force is applied during the scraping procedure. Consequently, most procedural failures can be traced back to low cell yields during this harvest step. Typically, cells are lost owing to physical destruction via over aggressive scraping. Alternatively, ineffective scraping may leave the majority of cells attached to the plates, which are then discarded. Therefore, it is essential to monitor the harvest procedure closely by visual inspection of the plates. Finally, the scraping procedure can be quite time consuming. Therefore, it is also imperative to ensure that the pH of the culture medium does not become excessively alkaline during this process, which may reduce cell viability. Therefore, only a few plates should be harvested at one time until some level of proficiency is obtained.
10. Alternatively, if one desires to recover the immunodepleted cell populations for further analysis the immunodepletion can be performed using the CELLection™ Biotin Binder Kit (Dynal Biotech). Herein, biotinylated antibodies are attached to CELLection Dynabeads via a DNA linker, which provides a cleavable site to remove the beads from the cells.
11. Often people inquire whether we have used automated systems to purify MSCs directly from bone marrow, e.g., MACS (Miltenyi Biotec). We do not anticipate this method to be effective for the following reason. According to the literature the frequency of MSCs in bone marrow ranges form 1 cell in 10^5–10^6 marrow cells. Because we typically obtain approximately 1×10^9 marrow cells from 40 donor animals, separation at this stage would yield 10^3–10^4 MSCs. This anticipated low yield is the main reason why plastic adherent populations are expanded ex vivo as a first step toward enrichment of MSCs.
12. Antibody conjugated Dynabeads can be prepared one day before use and stored overnight at 4 °C.
13. A distinguishing characteristic of IDmMSCs is their slow growth in vitro. Populations undergo only one or two population doublings following the first week of culture after immunodepletion. At present, the reason for their reduced growth potential remains under investigation. However, it is clear that cross-talk with hematopoietic cell lineages regulates expansion of the stromal cell layer.
14. The yield of IDmMSCs varies significantly between different inbred strains of mice *(12)*. These differences, in part, parallel differences in bone mineral density between strains. We prefer to use the FVB/n as marrow donors owing to their high yield of IDmMSCs.

15. Based on our experience, freshly isolated IDmMSCs exhibit poor viability when subjected to sheer forces during analysis by flow cytometry or fluorescent-activated cell sorting (FACS). Consequently, we typically culture IDmMSCs for 4–5 d post-immunodepletion before analyzing their cell surface phenotype by flow cytometry or FACS. Reduced viability is likely due to the repeated and prolonged manipulation of cells during the immunodepletion process. Similarly, we typically culture IDmMSCs for 5–7 d in complete culture media before analysis of their differentiation potential even though their differentiation capacity is not significantly altered by the immunodepletion process.
16. IDmMSCs are harvested by washing the cell monolayer with serum-free α-MEM, incubating with trypsin-EDTA (0.25%) for 3–5 min at room temperature and collecting the cells by washing repeatedly with medium. Physical scraping is not required to harvest IDmMSCs.
17. When staining IDmMSCs (or any mouse cells) with antibodies generated in mice it is essential to block low affinity Fc-mediated binding of antibodies to the mouse FC receptor. This is done by incubating cells for several minutes with the Mouse Fc-block™ (anti-mouse CD16/CD32 monoclonal antibody). This assures that any observed staining is owing to the interaction of the antigen-binding portion of the antibody with an antigen on the cell surface.
18. Another distinguishing feature of IDmMSCs is their limited capacity for adipogenic differentiation. This contrasts with immortalized stromal cells lines, which typically show enhanced adipogenic differentiation and reduced or absent chondrogenic potential. Therefore, although adipocytes become visible after 4–7 d post-induction only a small fraction of the total cell population (<20%) will differentiate into adipocytes using the methods we describe. We have not attempted to optimize the adipogenic induction protocol to increase the frequency of lineage commitment within populations.
19. Suspending IDmMSCs in chondrogenic induction medium before centrifugation eliminates the need to change the medium immediately after formation of the micromass pellet. After several days in the culture the pellets are less fragile thereby facilitating subsequent medium changes.
20. The micromass pellet increases in size owing to the secretion by chondrocytes of a large amount of proteoglycans into the extracellular space. Furthermore, these proteoglycans are hydrophilic and therefore have a high water content (recall that cartilage is mostly water, which enhances its shock absorbing capacity). Lack of detectable changes in pellet size may indicate a failure of cells to differentiate.
21. Safranin O is typically preferred over Toluidine blue for staining cartilage owing to its linear saturation response with increasing proteoglycan content. Accordingly, it provides a more accurate measure of the extent of chondrocyte differentiation.
22. The optimal plating density of IDmMSCs for osteogenic differentiation should be determined empirically. Plating density must be high enough to promote osteogenic commitment but also allow for expansion of the osteoprogenitor cells during the differentiation time course. If cells are plated at too high a density the cell monolayer may become so dense that it detaches from the culture dish and rolls-up into an insoluble aggregate.

Acknowledgments The author would like to thank Maria DuTreil and Dina Gaupp for assistance in preparing this chapter.

References

1. Phinney, D. G. (2002) Building a consensus regarding the nature and origin of mesenchymal stem cells. *J. Cell. Biochem.* **Suppl 38**, 7–12.
2. Dennis, J. E., Merriam, A., Awadallah, A., Yoo, J. U., Johnstone, B., and Caplan, A. I. (1999) A quadripotential mesenchymal progenitor cell isolated from the marrow of an adult mouse. *J. Bone Min. Res.* **14**, 700–709.

3. Pittenger, M. F., MacKay, A. M., Beck, S. C., Jaiswal, R. K., Douglas, R., Mosca, J. D., et al. (1999) Multilineage potential of adult human mesenchymal stem cells. *Science* **284**, 143–147.
4. Mauraglia, A., Cancedda, R., and Quarto, R. (2000) Clonal mesenchymal progenitors from human bone marrow differentiate in vitro according to a hierarchical model. *J. Cell Sci.* **113**, 1161–1166.
5. Kuznetsov, S. A., Krebsbach, P. H., Satomura, K., Kerr, J., Riminucci, M., Benayahu, D., et al. (1997) Single-colony derived strains of human marrow stromal fibroblasts form bone after transplantation in vivo. *J. Bone Min. Res.* **12**, 1335–1347.
6. Caplan, A. I. (1994) The mesengenic process. *Clin. Plastic Surg.* **21**, 429–435.
7. Bearpark, A. D. and Gordon, M. Y. (1989) Adhesive properties distinguish sub-populations of hematopoietic stem cells with different spleen colony-forming and marrow repopulating capacities. *Bone Marrow Transpl.* **4**, 625–628.
8. Kerk, D. K., Henry, E. A., Eaves, A. C., and Eaves, C. J. (1985) Two classes of primitive pluripotent hematopoietic progenitor cells: Separation by adherence. *J. Cell Physiol.* **125**, 127–134.
9. Deryugina, E. I. and Muller-Sieburg, C. E. (1993) Stromal cells in long-term cultures: Keys to the elucidation of hematopoietic development. *Crit. Rev. Immun.* **13**, 115–150.
10. Simmons, P. J., Masinovsky, B., Longenecker, B. M., Berenson, R., Torok-Storb, B., and Gallatin, W. M. (1992). Vascular cell adhesion molecule-1 expressed by bone marrow stromal cells mediates the binding of hematopoietic progenitor cells. *Blood* **80**, 388–395.
11. Witte, P. L., Robinson, M., Henley, A., Low, M. G., Stiers, D. L., Perkins, S., et al. (1987) Relationship between B-lineage lymphocytes and stromal cells in long-term bone marrow cultures. *Eur. J. Immunol.* **17**, 1473–1484.
12. Phinney, D. G., Kopen, G., Isaacson, R. L., and Prockop, D. J. (1999) Plastic adherent stromal cells from the bone marrow of commonly used strains of inbred mice: Variations in yield, growth, and differentiation. *J. Cell. Biochem.* **72**, 570–585.
13. Gordon, M. Y., Bearpark, A. D., Clarke, D., and Dowding, C. R. (1990) Haemopoietic stem cell subpopulations in mouse and man: Discrimination by differential adherence and marrow repopulating ability. *Bone Marrow Transpl.* **5(Suppl)**, 6–8.
14. Kiefer, F., Wagner, E. F., and Keller, G. Fractionation of mouse bone marrow by adherence separates primitive hematopoietic stem cells from in vitro colony-forming cells and spleen colony-forming cells. *Blood* **78**, 2577–2582.
15. Winemann, J. P., Nishikawa, S-I., and Muller-Sieburg, C. E. (1993) Maintenance of high levels of pluripotent hematopoietic stem cells in vitro: effect of stromal cells and c-kit. *Blood* **81**, 365–372.
16. Wang, Q-R. and Wolf, N. S. (1990) Dissecting the hematopoietic microenvironment. VIII. Clonal isolation and identification of cell types in murine CFU-F colonies by limiting dilution. *Exp. Hematol.* **18**, 355–359.
17. Modderman, W. E., Vrijheid-Lammers, T., Lowik, C. W., and Nijweide, P. J. (1994) Removal of hematopoietic cells and macrophages from mouse bone marrow cultures: isolation of fibroblast-like stromal cells. *Expt. Hematol.* **22**, 194–201.
18. Van Vlasselaer, P., Falla, N., Snoeck, H., and Mathieu, E. (1994) Characterization and purification of osteogenic cells from murine bone marrow by two-color cell sorting using anti-Sca-1 monoclonal antibody and wheat germ agglutinin. *Blood* **84**, 753–763.
19. Kopen, G., Prockop, D. J., and Phinney, D. G. (1999) Marrow stromal cells migrate throughout forebrain and cerebellum, and they differentiate into astrocytes after injection into neonatal mouse brains. *Proc. Natl. Acad. Sci. USA*. **96**, 10,711–10,716.
20. Baddoo, M., Hill, K., Wilkinson, R., Gaupp, D., Hughes, C., Kopen G. C., et al. (2003) Characterization of mesenchymal stem cells isolated from murine bone marrow by negative selection. *J. Cell. Biochem.* **89**, 1235–1249.
21. Phinney, D. G., Hill, K., Michelson, C., Dutreil, M., Hughes, C., Humphries, S., et al. (2006) Biological activities encoded by the murine mesenchymal stem cell transcriptome provide a basis for their developmental plasticity and broad clinical efficacy. *Stem Cells* **24**, 186–198.

22. Meirelles, L. S. and Nardi, N. B. (2003) Murine marrow-derived mesenchymal stem cell: isolation, in vitro expansion and characterization. *Brit. J. Hematol.* **123**, 702–712.
23. Sun, S., Guo, Z., Xiao, X., Liu, B., Liu, X., Tang, P-H., et al. (2003) Isolation of mouse marrow mesenchymal progenitors by a novel and reliable method. *Stem Cells* **21**, 527–535.
24. Peister, A., Mellad, J. A., Larson, B. L., Hall, B. M., Gibson, L. F., and Prockop, D. J. (2004) Adult stem cells from bone marrow (MSCs) isolated from different strains of inbred mice vary in surface epitopes, rates of proliferation, and differentiation potential. *Blood* **103**, 1662–1668.
25. Prowse, K. R. and Greider, C. W. (1995) Developmental and tissue-specific regulation of mouse telomerase and telomere length. *Proc. Natl. Acad. Sci. USA* **92**, 4818–4822.
26. Dobson, K. R., Reading, L., Haberey, M., Marine, X., and Scutt, A. (1999) Centrifugal isolation of bone marrow from bone: an improved method for the recovery and quantification of bone marrow osteoprogenitor cells from rat tibiae and femurae. *Calc. Tissue Internat.* **65**, 411–413.
27. Johnstone, B, Hering, T. M., Caplan, A. I., Goldberg, V. M., and Yoo, J. U. (1998) In vitro chondrogenesis of bone marrow-derived mesenchymal progenitor cells. *Exp. Cell Res.* **238**, 265–272.
28. MacKay, A. M., Beck, S. C., Murphy, J. M., Barry, F. P., Chichester, C. O., and Pittenger, M. F. (1998) Chondrogenic differentiation of cultured human mesenchymal stem cells from marrow. *Tissue Eng.* **4**, 415–428.
29. Gregory, C. A., Gunn, W. G., Peister, A., and Prockop, D. J. (2004) An Alizarin red-based assay of mineralization by adherent cells in culture: comparison with cetylpyridinium chloride extraction. *Anal. Biochem.* **329**, 77–84.

Index

MIX
Papier aus verantwortungsvollen Quellen
Paper from responsible sources
FSC® C105338

If you have any concerns about our products, you can contact us on
ProductSafety@springernature.com

In case Publisher is established outside the EU, the EU authorized representative is:
Springer Nature Customer Service Center GmbH
Europaplatz 3, 69115 Heidelberg, Germany

Printed by Libri Plureos GmbH
in Hamburg, Germany